Materials Horizons: From Nature to Nanomaterials

Series Editor

Vijay Kumar Thakur, School of Aerospace, Transport and Manufacturing, Cranfield University, Cranfield, UK

Materials are an indispensable part of human civilization since the inception of life on earth. With the passage of time, innumerable new materials have been explored as well as developed and the search for new innovative materials continues briskly. Keeping in mind the immense perspectives of various classes of materials, this series aims at providing a comprehensive collection of works across the breadth of materials research at cutting-edge interface of materials science with physics, chemistry, biology and engineering.

This series covers a galaxy of materials ranging from natural materials to nanomaterials. Some of the topics include but not limited to: biological materials, biomimetic materials, ceramics, composites, coatings, functional materials, glasses, inorganic materials, inorganic-organic hybrids, metals, membranes, magnetic materials, manufacturing of materials, nanomaterials, organic materials and pigments to name a few. The series provides most timely and comprehensive information on advanced synthesis, processing, characterization, manufacturing and applications in a broad range of interdisciplinary fields in science, engineering and technology.

This series accepts both authored and edited works, including textbooks, monographs, reference works, and professional books. The books in this series will provide a deep insight into the state-of-art of Materials Horizons and serve students, academic, government and industrial scientists involved in all aspects of materials research.

Review Process

The proposal for each volume is reviewed by the following:

1. Responsible (in-house) editor
2. One external subject expert
3. One of the editorial board members.

The chapters in each volume are individually reviewed single blind by expert reviewers and the volume editor.

Manish Kumar · Samiksha Dabas

Multiferroic and Magnetoelectric Materials

Applications and Emerging Green Technologies

Springer

Manish Kumar
Experimental Research Laboratory
Department of Physics, ARSD College
University of Delhi
New Delhi, Delhi, India

Samiksha Dabas
Department of Physics
Netaji Subhas University of Technology
New Delhi, Delhi, India

ISSN 2524-5384 ISSN 2524-5392 (electronic)
Materials Horizons: From Nature to Nanomaterials
ISBN 978-981-95-4390-8 ISBN 978-981-95-4391-5 (eBook)
https://doi.org/10.1007/978-981-95-4391-5

This Springer imprint is published by the registered company Springer Nature Singapore Pte Ltd.
The registered company address is: 152 Beach Road, #21-01/04 Gateway East, Singapore 189721, Singapore

“Dedicated to our Beloved Parents”

Preface

We are pleased to introduce this book, entitled *Multiferroic and Magnetoelectric Materials: Applications and Emerging Green Technologies*, to summarize the outstanding research and advanced technological developments in the field of magnetoelectric and multiferroic materials. This book covers the latest and greener technological applications, which is ranging from newer magnetoelectric and multiferroic materials to their technological aspects and provide an insight into their fundamental properties, synthesis methods, characterisation techniques, current uses and potential future applications. The book commences with an introduction to definition and need for magnetoelectric and multiferroic materials and an outline of basic concepts related to ferroelectricity and magnetism. This is followed by need and types of magnetoelectric and multiferroic materials, mechanisms associated with magnetoelectric coupling, recent advancement in technologies, applications and device fabrication. The unique characteristics and processes that allow these materials to possess specific value of ME coupling are discussed in-depth. The fabrication methods, basic characterisation techniques, green energy, photocatalytic and energy harvesting applications, spintronic device technology and applications concerned with direct and indirect magnetoelectric coupling are also thoroughly discussed.

As per the requirement of the multiferroic and magnetoelectric materials and associated multifunctional emerging device technologies, this book is divided into the seven dedicated chapters, namely:

Chapter 1, "Multiferroic and Magnetoelectric Materials: Fundamentals and Properties" highlights the historical origin of multiferroic and magnetoelectric materials, basic concepts of dielectric relaxation, ferroelectric and magnetic properties, exchange interactions. This chapter also covers types of multiferroics, magnetoelectric materials and magnetoelectric coupling. Finally, the chapter concludes with holistic view on recent breakthroughs in technologies, applications and device fabrication.

Chapter 2 "Synthesis Techniques and Measurements for Various Physical Properties of Multiferroic and Magnetoelectric Materials" emphasizes on synthesis techniques such as solid-state reaction, hydrothermal, sol-gel, molecular beam epitaxy, chemical vapour deposition, pulsed laser deposition, floating zone and green synthesis. Further, a comprehensive discussion on measurement techniques i.e., dielectric spectroscopy, ferroelectric and magnetic measurements, impedance spectroscopy and magnetoelectric coupling measurement methods is provided.

Chapter 3 "Sustainable Multiferroic and Magnetoelectric Materials for Energy Harvesting Applications" explores the photocatalytic properties, portable electronic devices, energy harvesting devices. This chapter concludes with a discussion on the synergy between magnetoelectric and photocatalytic strategies.

Chapter 4 "Photovoltaic Performance of Sustainable Multiferroic and Magnetoelectric Materials in Green Energy Technology" reviews the theory on Shockley-Queisser limit, quantum efficiency and applications in multiferroic and magnetoelectric materials. It provides insights into green energy applications and advancements in the material synthesis.

Chapter 5 "Magnetoelectric Multiferroics for Spintronic Applications" elucidates the core concepts of spin manipulation and spin-dependent conduction mechanism. This chapter delves into magnetoresistance including giant magnetoresistance, spin valve structures, tunnelling magnetoresistance, multiferroic tunnel junctions, spin current, Spin Hall Effect and its applications in bilinear magnetoelectric resistance. Further, this chapter provides insight into spintronic devices including magnetoelectric spin-orbit device and magnetoelectric spin-field-effect transistor multiplexer.

Chapter 6 "Advances in Magnetoelectric Devices and Applications in Surface Acoustic Wave Device Technology" highlights the latest developments in magnetoelectric sensor technology, including in magnetic field sensor device and related biomedical applications, current sensor, magnetoelectric gyrator, magnetoelectric transformer, wireless power transfer, magnetoelectric random access memory, microwave tunable filter devices, phase shifters, magnetoelectric voltage tunable inductors. It also elucidates the surface acoustic waves and the role of strain mediated magnetoelectric coupling. Finally, the chapter concludes with device applications including acoustic driven magnetic resonance and skyrmions.

Chapter 7 "Conclusion and Future Outlook on Sustainable Multiferroic and Magnetoelectric Material-Based Technologies" summarizes the summary of the book and encompass review on ongoing research to overcome the challenges associated with materials, device applications and reduced environmental impact. It also focusses on potential direction for the future research work.

In conclusion, the present book provides a comprehensive outline on multiferroic and magnetoelectric materials and related recent advancements in technology. It aims to strengthen the understanding and lay the groundwork for future advancements in low power consuming portable electronics.

In addition to the industry and academia, we believe that this book will be very helpful to undergraduate, graduate, and research scholars working in the fields of materials science, physics, chemistry, and many interdisciplinary engineering departments.

New Delhi, India

Dr. Manish Kumar
Dr. Samiksha Dabas

Acknowledgements We take great pride in expressing our appreciation to everyone who has helped us along the way and contributed to the creation of this book. We must first acknowledge and give thanks to "GOD" for his favours, which include our protection and good health throughout this time. Without our faith in the almighty, we could never have achieved this arduous task.

The authors extend the deepest gratitude to Prof. Gyantosh Kumar Jha, Principal, ARSD College, University of Delhi, New Delhi, India and H'ble Vice-Chancellor Prof. Anand Srivastava, Prof. Vinod Kumar, Head, Department of Physics, Netaji Subhas University of Technology, New Delhi, India for their unwavering support. Our special thanks to Prof. O. P. Thakur Department of Physics, Netaji Subhas University of Technology, New Delhi, India for his invaluable encouragement and suggestions throughout the completion of this book.

The authors would also like to thank the entire Springer Nature Editorial team for their unwavering initiative and diligent effort throughout the book's publishing process. Furthermore, we would like to continue by expressing our gratitude to the production team and their employees for their support and help, which was made possible by their dedication, skill, and pursuit of greatness.

We want to acknowledge the sincere thanks to our students specially, Ms. Khushboo Gupta, Mr. Manish Kumar and Mrs. Meena Devi for their unwavering support throughout this work. We would also like to extend our sincere gratitude towards our friends, family, and colleagues for their steadfast understanding, inspiration, valuable inputs and support along this journey.

Dr. Manish Kumar is also thankful to his grandparents and parents (Late Shri K. L. Vats (Grandfather), Late Smt. Shimla Devi (Grandmother), Shri Pravind Kumar (Father), Smt. Anita Sharma (Mother) for the endless support and understanding of my goals and aspirations. Their constant words of inspiration and confidence in me have been a source of encouragement and enthusiasm. I want to specially mention the names of my lovely wife Tanu and cute daughter Vanya for the all-time support and understanding towards my work without any kind of complain. I am really very thankful to both of my life lines. I also want to owe my deepest gratitude towards my uncle and aunty, Shri Mithelesh Sharma, Smt. Suman Sharma, Shri Sanjay Sharma and Smt. Sarika Sharma, cute and caring sisters, Shikha and Reshu, cousin brothers Ankit, Priyanshu and Upanshu, nephews and niece, Vidit, Shivohm, Vinayak and Pihu for support and encouragement in any situation.

Dr. Samiksha Dabas is thankful to her grandparents and parents (Shri Surajmal Dabas (Grandfather), Late Smt. Manhori devi (Grandmother), Shri Jawahar Singh Dabas (Father), Smt. Kamlesh Dabas (Mother). Besides, I owe my deepest gratitude to all my family members and friends who have been a constant source of strength and provided their eternal support, encouragement and understanding in all situations.

Additionally, we would like to thank everyone who helped to build and publish this book, whether directly or indirectly. Your help, encouragement, and dedication have made this endeavour incredibly valuable and fulfilling.

New Delhi, India

Dr. Manish Kumar
Dr. Samiksha Dabas

Competing Interests The authors have no competing interests to declare that are relevant to the content of this manuscript.

Contents

About the Authors

Dr. Manish Kumar (Corresponding Author) obtained his M.Tech. degree in Ceramic Engineering from Indian Institute of Technology, Banaras Hindu University, Varanasi and Ph.D. degree in Physics from Institute of Science, Department of Physics, Banaras Hindu University, Varanasi, India. At present, he is serving as **Assistant Professor** at the Department of Physics, ARSD College, University of Delhi, New Delhi, India for more than 12 years. His research area focuses on environmental friendly multifunctional materials for energy and memory devices, materials for green energy solutions, multiferroics, magnetoelectrics, composites, perovskite materials for photovoltaic applications, supercapacitors, white light emitting materials, solar simulation via different softwares and DFT calculation. **Recently, he is listed among top 2% scientists worldwide by Stanford University Ranking-2021, 2025** and published more than **135 research publications** in international journals of high repute, **one authored book on White Light Emitting Materials** and **five edited books** (in press) in Springer Nature, **30 international book chapters** and **one Indian Patent** with **h-index of 30, i10-index of 74** including more than **2900 citations** as per Google-Scholar. He has delivered the talks in a number of national and international conferences. In addition, he has completed two research projects as co-investigator funded by University of Delhi. He is guiding a number of Ph.D., M.Tech./M.Sc. and B.Sc./B.Tech. students as supervisor and/ co-supervisor. Dr. Kumar is serving/served as an **Editor/Guest Editor in** a number of highly reputed journals such as, **Scientific Reports (Nature)**, **Materials Today Communications (Elsevier),** ECS Sensors Plus (IOP, ECS), High Performance Polymers (SAGE), Materials Today: Proceedings (Elsevier) and Springer Nature Books. He has organized one national conference (RAFM-2020) as co-convener and three international conferences (RAFM-2022, RAFM-2024 and APGC-2025) as convener.

Dr. Samiksha Dabas (Co-Author) is an Assistant Professor in the Department of Physics, Netaji Subhas University of Technology, New Delhi, India. She completed her M.Sc. and also Ph.D. in Physics from University of Delhi. The fundamental physics of novel functional ferrite materials has seen a phenomenal rise owing to their multifarious applications viz., biomedical, automotive and electronic industry, microelectronics, gas sensors, thermal sensing switches, MW absorbers and so on. In this respect, her research for 'designer and new materials' is characterized by specific functionalities obtained as a result of theoretical and/or first principles simulations, experimental results and applicability to existing materials and methods. Specific materials of research interest are the magnetoelectric multiferroic materials whose applications include energy storage and harvesting, spintronics, sensors and memory devices for next generation multifunctional devices. Further, her research domain includes electromagnetic interference shielding applications utilizing materials based on ferrites-conducting polymers/MWCNT/MXenes. She is the reviewer of various national and international research journals. Besides, she has published many research publications in international journals of high repute for which she has also received "commendable research award". She has presented work and delivered invited talks in many national and international conferences. She has also published an Indian patent on "Heterostructure device for harnessing solar energy". She has been currently guiding a number of M.Sc. dissertation and Ph.D. students as supervisor and/or co-supervisor.

Chapter 1
Multiferroic and Magnetoelectric Materials: Fundamentals and Properties

1.1 Introduction

The interlinkage between electricity and magnetism was corroborated through Oersted experiment in 1820 and Faraday's experiments in 1831. Subsequently, the inter-dependence of the magnetic and electric field was formulated and finally assembled by Maxwell in the 1860s in the form of four Maxwell's differential equations. Pierre Curie assumed the likelihood of controlling polarization using the magnetic field and magnetization by the electric field. Also, a remark on the possible existence of the coupling between the two i.e., magnetic and electric orders was made by Landau and Lifshiz [1]. Existence of magnetoelectricity of Cr_2O_3 was predicted by Dzyaloshinskii in 1960 and subsequently measured by Astrov in 1961 [2, 3]. The term, MFs was however, coined by Schmid in 1994 [4]. With the further technological advances, the researchers further attempted to incorporate ferroelectricity and magnetism into a single-phase material that subsequently came to be known as MF. However, today MFs have much broader definition and also include the AFM order [5, 6].

Both the orders i.e., FE and magnetic have different origin in the perovskite structures [7]. In the perovskites such as $BaTiO_3$ and $Pb(ZrTiO_3)$, the FE behavior arises due to the transition ion such as Ti^{4+} whose empty d^o orbital form bond with p orbital of O^{2-} ($2p^6$) thereby, resulting in a shift of Ti^{4+} ionic position. However, the magnetic moment arises from the d orbital electrons only. So, the possibility of off-center shift that gives rise to ferroelectricity is subdued when the transition metal ion possesses d electrons. Thus, both the orderings i.e., FE and magnetic become mutually exclusive of each other. This is also shown in Fig. 1.1. This contradiction prompted the researchers to realize the alternative pathways including, (i) the materials in which ferroelectricity arises due to the processes other than the displacement mechanism of transition ions and, (ii) congruous with the magnetic ordering. A major development occurred in 2003 onwards with the materials such as $TbMn_2O_5$, $TbMnO_3$ (perovskite manganite) whereby it was observed that the magnetic field induced an

M. Kumar and S. Dabas, *Multiferroic and Magnetoelectric Materials*, Materials Horizons: From Nature to Nanomaterials, https://doi.org/10.1007/978-981-95-4391-5_1

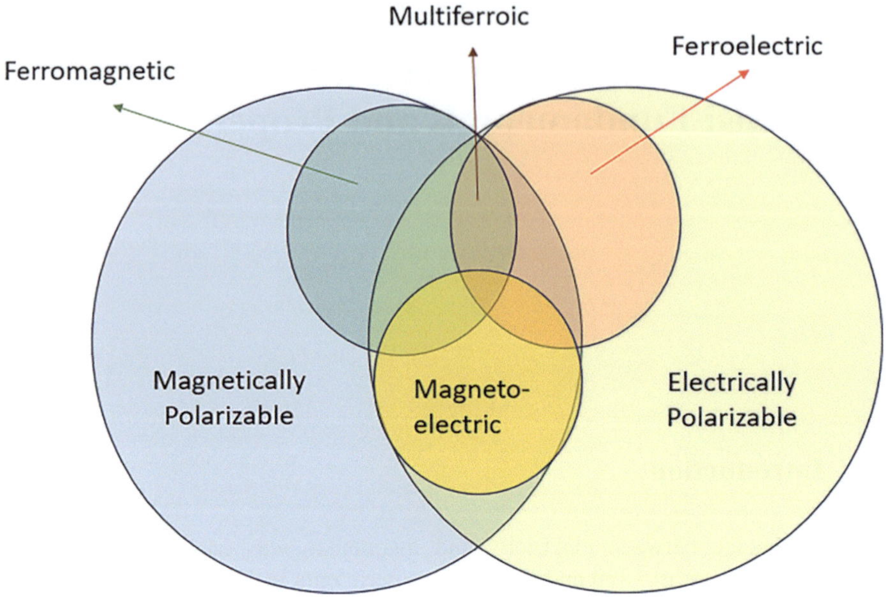

Fig. 1.1 Representation of the MF materials as a subclass of other materials

electric polarization [8, 9]. In these materials, a spiral AFM ordering occurs due to spin frustration. In addition, $TbMnO_3$ does not possess the s-orbital lone pairs as was the case in $BiFeO_3$ (ferroelectricity on account of lone pair mechanism). This has revitalized and accelerated the research interest in the field of MFs.

Certainly, technology advancement has effectuated the use of multifunctional devices that couple both electric as well as magnetic properties in a device [20]. Use of ferroelectrics include FE i.e., tunable capacitor, phase shifter, FERAM i.e., FeRAM, pyroelectric and piezoelectric sensor devices, etc. Alternatively, the FM materials find applications in data recording and storage such as in magnetic tapes, hard disk. With the combination of the two properties, the ME device can be utilized in the field of energy harvesters, highly sensitive sensors, spintronic, filters, etc. [10-14]. For microelectronic and sensor applications, MFs require to possess ME coupling of more than 0.1 V cm^{-1} Oe^{-1} at temperatures higher than room temperature, and low conductivity (less than 10^{-11} Ω cm) [15]. The MF $BiFeO_3$ suffice the high temperature requirement but possess low ME coupling coefficient in bulk form. This issue was resolved in early 2000s when epitaxial thin films of $BiFeO_3$ were made possessing ME coupling coefficient of 3 V cm^{-1} Oe^{-1} [16]. Further, $BiFeO_3$–$BaTiO_3$ based current sensor can have sensitivity 4.5 mV/A (error ~5%) in the frequency range from 1 to 20 kHz [17]. The ME coupling coefficient of 0.87 mV cm^{-1} Oe^{-1} has been observed in $BaTiO_3$–0.725$BiFeO_3$ ceramics [18]. Also, the room temperature ME coupling coefficient of 515 mV cm^{-1} Oe^{-1} has been observed in $BaTiO_3$/$BiFeO_3$/$BaTiO_3$ thin film heterostructures [19]. Reportedly, high energy storage density values up to 150 J/cm^3 have been observed for rare earth doped

$BiFeO_3$ films [21]. In addition, the energy density of 70 J/cm^3 has been observed for $BiFeO_3$–$SrTiO_3$ films [22].

A comprehensive description of basic concepts of dielectric relaxation, FE polarization followed by magnetism is elucidated. A Brief background description about the exchange interactions is also provided. Further, the basic introductory concepts about MFs, its types and ME coupling analysis are outlined. The analysis pertaining to doping and composite formation is also presented. The chapter ends with a brief overview of $BiFeO_3$, $BaTiO_3$, $CoFe_2O_4$ and the solid solutions of $BiFeO_3$–$BaTiO_3$.

1.2 Effect of Electric Field: Polarization

Polarization is the phenomena of displacement of electric charges from their equilibrium position due to an applied electric field in a dielectric. The positive charge gets displaced in the direction of electric field and vice versa. Various types of polarization observed are frequency dependent and are shown in Fig. 1.2 and are briefly described below:

(i) **Electronic Polarization**: Herein, the electron cloud of the atom gets displaced with respect to its equilibrium position due to the application of electric field. It is observed in monatomic gases at high frequency or optical region i.e., 10^{15} Hz.
(ii) **Orientational Polarization**: Due to the applied electric field, the centers of positive and negative charges separate and the dipoles get aligned with respect to the electric field. It occurs in audio frequency i.e., 10^6–10^{10} Hz.
(iii) **Ionic Polarization**: The constituent ions in the lattice (such as NaCl) get displaced due to the applied electric field. It occurs in the infrared region i.e., 10^{13} Hz.

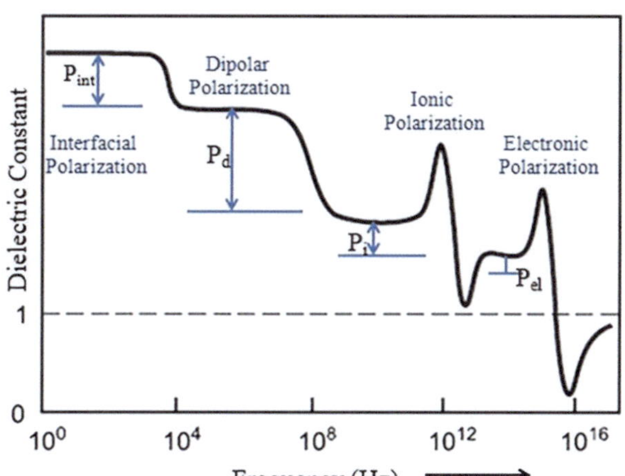

Fig. 1.2 Schematic representation of polarization types in frequency dependent dielectric plot

(iv) **Interfacial or Space Charge Polarization**: It occurs at the interface of the two heterogeneous faces or boundaries. Here, the charges get accumulated at the interface because of the difference of permittivity, conductivity, etc. which may be due to defects or cracks occurring inherently or during the material synthesis process. It occurs at the low frequencies i.e., $<10^2$ Hz.

1.3 Dielectric Relaxation

The change in the applied electric field does not instantaneously translate into change in the molecular polarization, but is rather observed with some momentary delay. This delay or lag is known as dielectric relaxation. Let the specimen be subjected to a constant or d.c electric field. Then, the dipoles will undergo orientational polarization, say, P_o. After the equilibrium reaches, the field is reduced to zero. So, the polarization will not instantly be zero instead, it will take some time i.e., relaxation time (τ). Then, the decay of polarization can be expressed as:

$$P_o(t) = P_o e^{-t/\tau}. \tag{1.1}$$

Interestingly, numerous theories have been utilized to explain the dielectric mechanism and properties of a material. These include Debye's theory and related ones like Cole-Davidson, Cole-Cole, Havriliak-Nigami. Also, the non-Debye type models include William-Watts, Jonscher model, etc. A brief background of Debye theory and Cole-Cole plots has been discussed here.

1.3.1 Debye Relaxation

Peter Debye gave the theory of Debye relaxation whereby, the dipoles take a characteristic time i.e., Debye relaxation time (τ), to get reoriented with respect to the changing applied electric field. It is assumed that the material possesses zero conductivity and comprises of non-interacting dipoles that follow Brownian motion and exhibit a single relaxation time. On the application of the sinusoidal electric field ($E_o\cos(\omega t)$), the expression for dielectric permittivity can be given in complex form, because of the phase difference in between orientational polarization ($P_0(\omega)$) and the applied electric field. Thus, we have:

$$\varepsilon^*(\omega) = \varepsilon' - j\varepsilon'', \tag{1.2}$$

where, ε' and ε'' are the real and imaginary part of complex dielectric permittivity. Both ε' and ε'' are frequency dependent and depicted in Fig. 1.3. Further analysis yields the following:

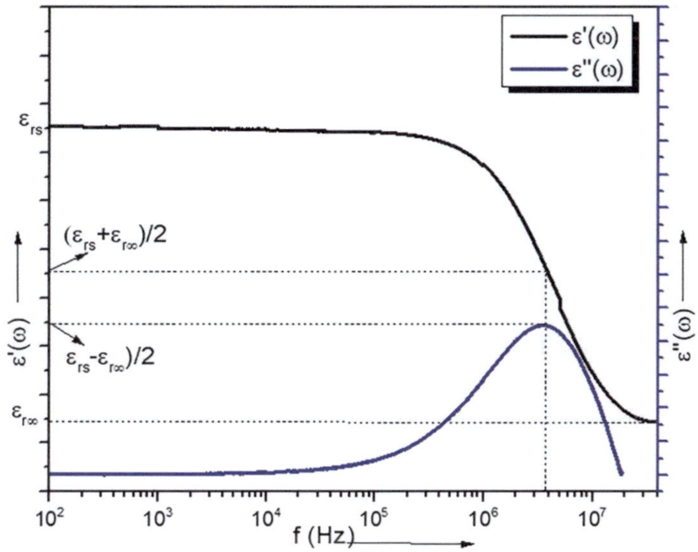

Fig. 1.3 Schematic of dielectric parameters as a function of frequency

$$\varepsilon^{*}(\omega) = \varepsilon_{\infty} + \frac{\varepsilon_s - \varepsilon_{\infty}}{1 - j\omega\tau}. \tag{1.3}$$

Notably, $\varepsilon_s - \varepsilon_{\infty}$ represents the strength of the relaxation mechanism. Here, $\omega \to 0$ (for constant electric field dielectric loss goes to zero) and permittivity i.e. $\varepsilon' \to \varepsilon_s$. Also, $\omega \to \infty \Rightarrow \varepsilon' \to \varepsilon_{\infty}$. Now, the real and imaginary part of dielectric permittivity can be obtained as follows:

$$\varepsilon'(\omega) = \varepsilon_{\infty} + \frac{\varepsilon_s - \varepsilon_{\infty}}{1 + \omega^2\tau^2}, \tag{1.4}$$

$$\varepsilon''(\omega) = \frac{\omega\tau(\varepsilon_s - \varepsilon_{\infty})}{1 + \omega^2\tau^2}. \tag{1.5}$$

These are known as the Debye equations. Furthermore, at a certain critical frequency i.e. $\omega_C\tau = 1$, we will obtain the maxima of the dielectric loss. Also,

$$\tan\delta = \frac{\text{Dielectric Loss}}{\text{Permittivity}} \text{ or } \frac{\varepsilon''}{\varepsilon'} = \frac{\omega\tau(\varepsilon_s - \varepsilon_{\infty})}{\varepsilon_s + \varepsilon_{\infty}\omega^2\tau^2} \to \omega\tau \text{ as } \omega \text{ approach zero.} \tag{1.6}$$

Subsequently, given a distribution function pertaining to the relaxation time i.e., $G(\tau)$, we can rewrite the relations given by Eqs. 1.4 and 1.5 as follows:

$$\varepsilon'(\omega) = \varepsilon_\infty + (\varepsilon_s - \varepsilon_\infty)\int_0^\infty \frac{G(\tau)}{1+(\omega\tau)^2}d\tau, \tag{1.7}$$

$$\varepsilon''(\omega) = (\varepsilon_s - \varepsilon_\infty)\int_0^\infty \frac{G(\tau)\omega\tau}{1+(\omega\tau)^2}d\tau. \tag{1.8}$$

1.3.2 Cole-Cole Plots

Evaluation of the relaxation time from the dielectric parameters utilizing Eqs. 1.7 and 1.8 is a tedious task. Accordingly, the modified-Debye relation was given by Cole-Cole as follows:

$$\varepsilon^*(\omega) = \varepsilon_\infty + \frac{\varepsilon_s - \varepsilon_\infty}{1+(j\omega\tau)^{1-\alpha}}. \tag{1.9}$$

Here, the value of the exponent parameter α lies in between 0 and 1. Then, the Debye equation can be obtained from a special case of $\alpha = 0$.

Also, using $j^{1-\alpha} = e^{j(1-\alpha)\frac{\pi}{2}}$ and simplifying further, the real and imaginary part of dielectric permittivity can be obtained as follows:

$$\varepsilon'(\omega) = \varepsilon_\infty + \frac{(\varepsilon_s - \varepsilon_\infty)\left[1+(\omega\tau)^{1-\alpha}\sin\left(\frac{\alpha\pi}{2}\right)\right]}{\left[1+2(\omega\tau)^{1-\alpha}\sin\left(\frac{\alpha\pi}{2}\right)+(\omega\tau)^{2(1-\alpha)}\right]}, \tag{1.10}$$

$$\varepsilon''(\omega) = \frac{(\varepsilon_s - \varepsilon_\infty)(\omega\tau)^{1-\alpha}\cos\left(\frac{\alpha\pi}{2}\right)}{\left[1+2(\omega\tau)^{1-\alpha}\sin\left(\frac{\alpha\pi}{2}\right)+(\omega\tau)^{2(1-\alpha)}\right]}. \tag{1.11}$$

Rearranging the terms, we get

$$\left(\varepsilon'(\omega) - \frac{(\varepsilon_s - \varepsilon_\infty)}{2}\right)^2 + \left(\varepsilon''(\omega) + \frac{(\varepsilon_s - \varepsilon_\infty)}{2}\tan\left(\frac{\alpha\pi}{2}\right)\right)^2 = \left[\frac{(\varepsilon_s - \varepsilon_\infty)}{2}\sec\left(\frac{\alpha\pi}{2}\right)\right]^2. \tag{1.12}$$

This equation is plotted in Fig. 1.4. This is the equation of a circle with center $\left(\frac{(\varepsilon_s-\varepsilon_\infty)}{2}, -\frac{(\varepsilon_s-\varepsilon_\infty)}{2}\tan\left(\frac{\alpha\pi}{2}\right)\right)$ that lies below x-axis and radius $\frac{(\varepsilon_s-\varepsilon_\infty)}{2}\sec\left(\frac{\alpha\pi}{2}\right)$.

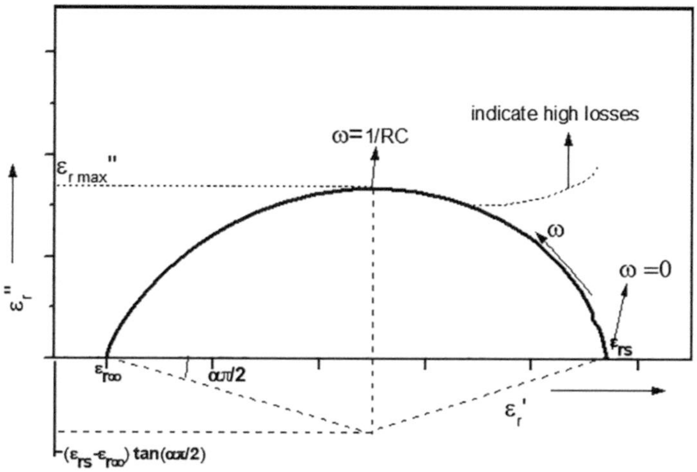

Fig. 1.4 Schematic of Cole-Cole plots

1.4 Ferroelectric Properties of Materials

FE properties for the first time were discovered by Joseph Valasek in 1921 in $KNaC_4H_4O_6H_2O$ or Rochelle salt [23]. Afterwards, Igor Kurchatov described ferroelectricity with a theoretical analysis [24]. Then, Busch and Scherrer described a FE namely, potassium dihydrogen phosphate in 1935 [25]. Subsequently, $BaTiO_3$, a FE oxide was discovered in the 1940s which accelerated the pace of research in the ferroelectrics [26, 28]. $BaTiO_3$ was the first FE without the presence of hydrogen bonds. In $BaTiO_3$, the ferroelectricity results from the sharing of electron cloud between oxygen (O^{2-}) and the transition ion (d^o configuration in Ti^{4+}).

In general, ferroelectrics are the materials which exhibit hysteresis i.e., the direction of FE polarization can be reversed on application of an alternating electric field. The FE hysteresis disappears above a critical temperature, known as FE Curie temperature (T_C), into a paraelectric state. Spontaneous polarization occurs when the material possess a dipole moment without the application of any electric field. Ferroelectrics are a subgroup of pyroelectric materials. As the applied alternating electric field increases, the domains get polarized as shown in the Fig. 1.5. After reaching saturation polarization (P_s), when the electric field is decreased, the domains start to back-switch. However, at zero electric field, there is still some net polarization i.e., P_r or remanent polarization. So, to get back to zero polarization, we need to reverse the direction of the applied electric field. This reversed field, where polarization is zero, is the coercive field (E_c). Subsequently, further increase in the field in the reverse direction will result in realignment of dipoles and finally a saturation polarization will be observed in the opposite direction. Finally, the applied electric field is reduced to zero value and this completes the hysteresis cycle.

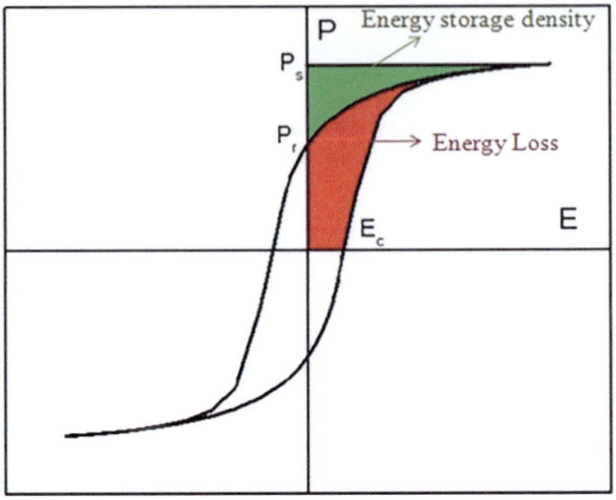

Fig. 1.5 Schematic of polarization versus electric field FE hysteresis

With the application of an electric field, the neighboring dipoles interact with each other resulting in a net electric dipole moment. Ferroelectrics obey Curie-Weiss law which is given as follows:

$$\frac{P}{\varepsilon_o E} = \frac{C}{T - T_C}, \tag{1.13}$$

where, T_C is the Curie temperature, C is Curie's constant. At T_C, we observe a FE to paraelectric phase transition. These transitions can be of (i) the first order whereby, we observe discontinuity in the latent heat, and (ii) the second order whereby, we observe discontinuity in the specific heat.

However, a few materials exhibit diffuse phase transitions (DPT) whereby, the phase transition, instead of being a peak anomaly, is diffused over a range of temperature or Curie range and is characterized by a gradual change in the phase. Thus, DPT is co-terminus with broadened dielectric permittivity maxima (dielectric versus temperature measurements), deviation with respect to the Curie-Weiss law in the transition region, the frequency dependence of dielectric parameters in the transition region, and a relaxation behavior in the phase transition region. Dielectric parameters in case of DPT can be understood based on the following relation [29]:

$$\frac{1}{\varepsilon} - \frac{1}{\varepsilon_m} = \frac{(T - T_m)^\gamma}{C}, \tag{1.14}$$

where, ε_m correspond to the maximum values of real part of permittivity at temperature T_m, C is a constant, γ is the critical exponent. The critical exponent has values in range $1 < \gamma < 2$, herein $\gamma = 1$ signifies Curie-Weiss based ferroelectric, and $\gamma = 2$ signifies a totally diffused phase transition.

1.5 Magnetic Properties of Materials

The origin of the magnetic properties lies in magnetic dipoles and is underpinned by two sources i.e., (i) orbital motion of the electron in an atom around the nucleus, and (ii) spin magnetic moment (spin up/down) [30]. Herein, with the equal and opposite spins, the net spin magnetic moment will be zero. Magnetic properties can be classified into diamagnetism, paramagnetism, Ferrimagnetism, ferromagnetism, and antiferromagnetism. This is shown diagrammatically in Fig. 1.6.

1.5.1 Ferromagnetism

In the FM materials, the spins of the unpaired or the valence shell electrons are aligned in a parallel position with respect to each other in a domain. The spins in parallel configuration will possess lesser value of exchange energy as compared with spins in antiparallel configuration. Ferromagnetism is governed by Curie-Weiss law and is given as:

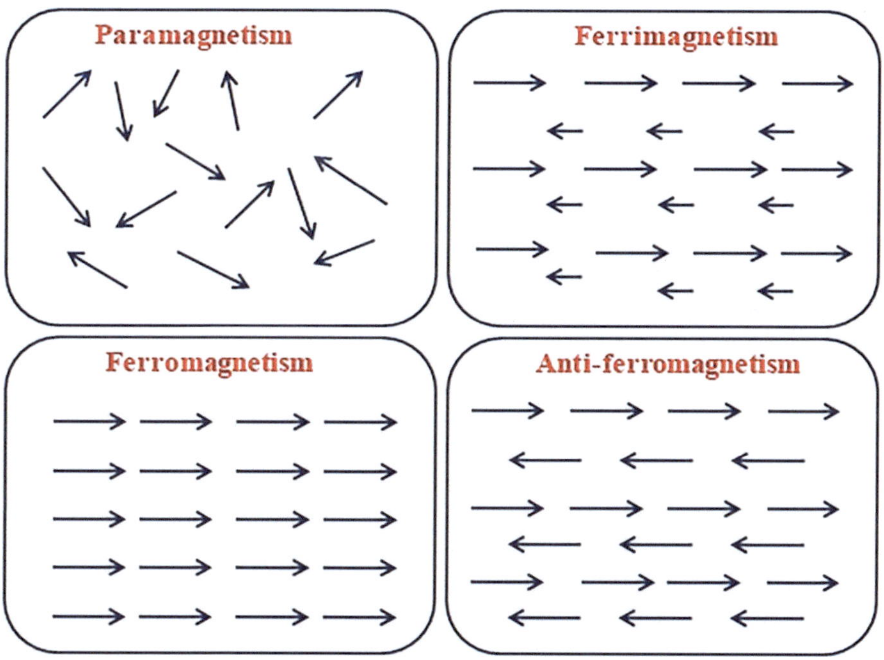

Fig. 1.6 Classification of magnetic orderings in magnetic materials

$$\chi = \frac{M}{H} = \frac{C}{T - \theta}, \tag{1.15}$$

where, χ is the susceptibility which is the ratio of M (magnetization) and H (applied magnetic field), θ is the Curie-Weiss temperature, and C is the Curie constant and is given by:

$$C = \frac{(g\mu_B)^2 S(S + 1)n}{3K_B}; \tag{1.16}$$

where, g is Lande's g factor, n is the cation density i.e., number of cations in a unit volume. FM materials, when subjected to an externally applied magnetic field, exhibit their characteristic hysteresis loop as shown in Fig. 1.7. With an increase in the applied magnetic field, the domains orient along direction of magnetic field, thus getting magnetized thereby showing an increase in magnetization value. Eventually, the domains get completely aligned with respect to the magnetic field thereby evincing saturation magnetization (Ms). Notably, above a critical threshold field, while reducing the magnetic field, magnetization of the material does not follow its initial path. Further, while demagnetizing the sample i.e., when the magnetic field goes to zero, the domains do not instantaneously switch back to their original position, and instead retain some remnant magnetization (M_r). Thereafter, a coercive field (H_c) is applied in opposite direction in order to obtain zero magnetization value. Eventually, we observe a complete hysteresis loop being traced out.

Notably, the changes in the size and the rotation of the domain walls due to the applied magnetic field can cause sudden and discrete jumps in magnetization value. This is referred to as Barkhausen noise whereby the inherent impurities/defects/dislocations can form a potential barrier the domain walls came across.

Further, magnetic materials can be classified in to two categories:

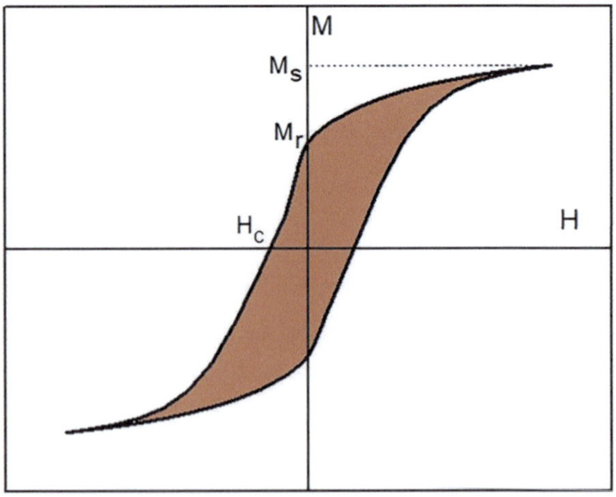

Fig. 1.7 Schematic of magnetization versus magnetic field hysteresis loop

(i) The **soft magnetic materials** are characterized by low coercivity, low permeability, low saturation magnetization, and less eddy current losses. On account of low coercivity, they can be readily magnetized and de-magnetized, and consequently, the energy dissipation is less. Uses include core of transformers, electromagnets and inductors. Examples of soft magnetic materials include Mn–Zn, Ni–Zn, Ni–Zn ferrite, etc.
(ii) The **hard magnetic materials** have large crystalline anisotropy and are characterized by high coercivity. They are relatively hard to magnetize and demagnetize. Uses include recording media, automobile applications, etc. Examples of hard magnetic materials include alloy Alnico, Co-Steel, $SmCo_5$, etc.

Exchange anisotropy or exchange bias originates in a material at the interface of two magnetic orders such as FM-AFM, which are coupled together such as in the case of Co/NiO, Co/CoO, etc. [31]. This can also include interface with respect to ferrimagnetic-AFM or FM-FM orders. Exchange anisotropy is observed when the material specimen is cooled (in static H) below the T_N of the AFM layer, whereby, the Curie temperature (T_C) of the FM interface is larger than T_N. Here, the shift is experienced by hysteresis loop opposite to the cooling field direction. This implies that the coercive field has different values with respect to the increasing and decreasing magnetic field.

1.5.2 Antiferromagnetism

In Antiferromagnetism, the materials possess sub-lattices with the spins corresponding to neighboring sub-lattices being aligned antiparallel or opposite to each other. Antiferromagnetism exists below Neel temperature. Magnetic susceptibility is given as

$$\chi = \frac{C}{T + \theta_N}. \tag{1.17}$$

Here, the material is known to undergo AFM to paramagnetic transition at Neel temperature (θ_N). Further, the AFM ordering in materials, as shown in Fig. 1.8, can be classified into 3 types namely,

(i) **A-type**: Herein, we have, FM intra-planar coupling along with the AFM inter-planar coupling.
(ii) **C-type**: Herein, we have, AFM intra-planar coupling along with the FM inter-planar coupling.
(iii) **G-type**: Herein, we have both the intra-planar and inter-planar coupling to be AFM.

The structure factor D for A-type (D_A) and G-type (D_G) AFM materials can be approximated as:

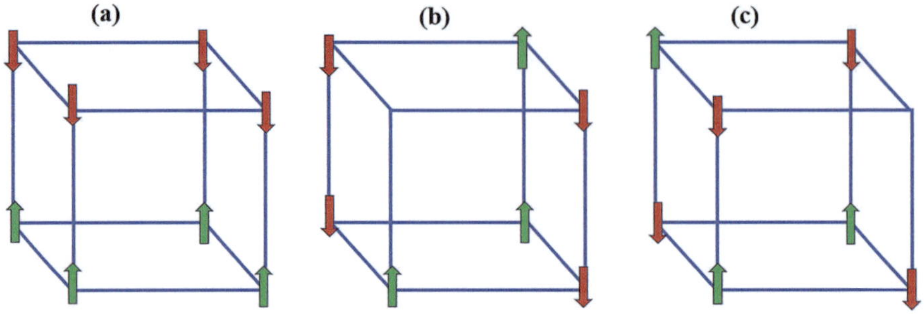

Fig. 1.8 Types of AFM ordering. **a** A-type, **b** C-type, and **c** G-type

$$D_A \sim \frac{1+S^2}{2}, \tag{1.18}$$

$$D_G \sim \frac{1+S^2}{4}, \tag{1.19}$$

where, S denotes the fraction of sites accurately located. Notably, for ferromagnets, D_F is proportional to S^2.

1.6 Exchange Interactions

In the transition metal oxide, the 'd' shell electron spins interact with the spins of the neighboring atom. This relative interaction is due to the electron spin configuration and results in the exchange interactions [32]. Exchange interactions are electrostatic. They are of three types (i) Direct exchange interaction, (ii) super exchange interaction, and (iii) Double exchange interaction.

1.6.1 Direct Exchange Interactions

These are short range interactions. The exchange energy (E_{ex}) between the two neighboring atoms can be expressed as:

$$E_{ex} = -2J_{ex}\, S_i \cdot S_j = -2J_{ex} S_i S_j \cos\theta, \tag{1.20}$$

where, S_i and S_j are the interacting spins with θ being the angle between them, and J_{ex} is the exchange integral. The positive sign of J_{ex} is associated with FM coupling, whereas the negative sign symbolizes AFM coupling. This is because with positive J_{ex}, minima of E_{ex} will correspond to $\theta = 0°$ or parallel electron spins orientation as

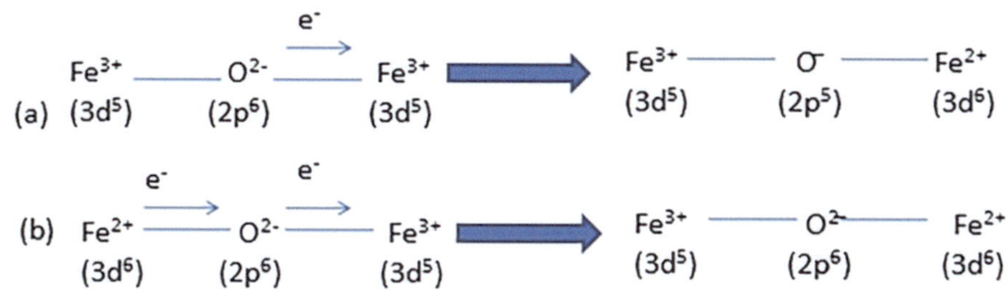

Fig. 1.9 Schematic for **a** superexchange and **b** double exchange interactions

in FM coupling. Similarly, in case of negative J_{ex}, minimum E_{ex} case will correspond to $\theta = 180°$ or antiparallel spins as in AFM coupling. J_{ex} also depends on the position and separation between the electron spins.

1.6.2 Superexchange Interactions

These are indirect exchange interactions characterized by the long range ordering. The interaction between the two magnetic ions (such as possessing partially filled d orbital shell) is effectively mediated by another ion which can be non-magnetic (such as oxygen ion in excited state) as explained in Fig. 1.9. It can be understood in the following way. The O^{2-} ($2p^6$) ion has fully filled p orbital and is in between the two Fe^{3+} ($3d^5$) ions having completely half-filled d orbital as per Hund's rule. Now, the superexchange interaction comes in to play. An electron from p orbital of oxygen ion leaves its p orbital site and in the process oxygen is now in its excited state i.e., O^- ion ($2p^5$). This electron briefly gets included in one of the Fe^{3+} ion and form Fe^{2+} ($3d^6$). Consequently, the unpaired electron present in O^- ion ($2p^5$) will interact with other Fe^{3+} ($3d^5$) ion.

$$H = \sum_{ij} J_{ij} S_i S_j, \tag{1.21}$$

where, J_{ij} is the exchange integral corresponding to super exchange interaction.

1.6.3 Double Exchange Interactions

The double exchange interactions were proposed by C. Zener and their origin lies in the different valence or oxidation states of ions such as Fe^{2+} and Fe^{3+} in Fe_3O_4 [33]. Here, there are two simultaneous processes. The O^{2-} ($2p^6$) ion is in between the Fe^{3+} ($3d^5$) and Fe^{2+} ($3d^6$) ions. As the unpaired electron from Fe^{2+} gets transferred to oxygen ion, simultaneously a p-orbital electron from oxygen ion gets transferred to Fe^{3+}.

1.6.4 Dzyaloshinkii-Moriya Interactions

Dzyaloshinkii-Moriya (DM) interactions refer to the antisymmetric exchange interactions whereby, a weak FM effect arises between two ions of spin S_i and S_j in an AFM material [34, 35]. The DM interaction energy can be expressed as:

$$E_{ij}^{\mathrm{DM}} = d_{ij}\cdot(S_i \times S_j), \tag{1.22}$$

where, d_{ij} represents the coupling vector. DM interaction causes a canting effect in magnetic moments thereby, resulting in a weak FM effect in an AFM material.

1.7 Multiferroics

MFs are defined as the set of materials known to possess two or more ferroic order parameters simultaneously such as ferroelectric, FM, and AFM orders. Some examples of MF materials are: Cr_2O_3, Gd_2CuO_4, $BiFeO_3$, $Ni_3B_7O_{13}I$, $Cr_3B_7O_{13}Cl$, $TbMn_2O_5$, $BaMF_4$ where M can be Co, Ni, Mn, Fe, etc. [36-39]. Harnessing the technological aspects of MFs is pivotal particularly in fields of FE random access memory (FeRAM), spintronics, magnetic field-based sensing applications, etc. Notably, single phase MFs are scarce and also it is difficult to obtain significantly high value of ME coupling coefficient in single phase MFs. Alternatives such as devising multi-layered heterogeneous composites or multi-phase MFs, doping appropriately are utilized whereby the material cumulatively exhibits MF behavior. The composites exhibiting multiferroicity via mechanisms that can be strain mediated, charge mediated, exchange bias mediated heterostructures and electron transport-based heterostructures [36, 40, 41]. The single phase MFs including $BiMnO_3$, $ReMnO_3$ (Re is rare earth like Tm, Yb, Ho, Er, Y, Sc, Lu), $ReMn_2O_5$ (Re can be Tb, Ho, Dy, Y) $BiFeO_3$ and $YMnO_3$, etc. $YMnO_3$ (hexagonal manganite) exhibit FE behavior as well as A-type antiferromagnetism. Figure 1.10 provides a brief representation of MFs.

1.7.1 Type-I Multiferroics

In the type-I MFs, the origin of the FEand the magnetic orders is separate i.e., both orders occur independently of each other. These typically exhibit weak magnetic electric coupling though; they generally possess good ferroelectricity [43-45]. The

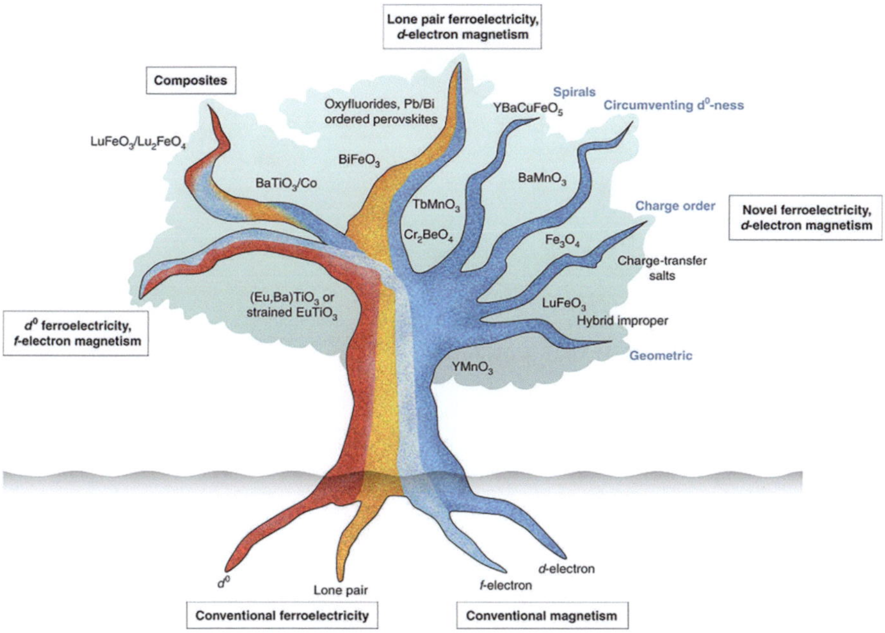

Fig. 1.10 Representation of multiferroicity underpinned by FE mechanism and magnetism in various materials. Adapted with permission from Ref. [6]

most extensively studied type 1 MF i.e., $BiFeO_3$ possesses ferroelectricity on account of the lone pairs present in Bi^{3+}. Few other examples are $BiMnO_3$, $YMnO_3$, $EuTiO_3$, $Ca_3Mn_2O_7$. In general, as shown in the Fig. 1.11, on the basis of FE mechanism in materials, type 1 MFs are classified into various sub-classes as follows:

(i) **MF perovskites**: Pervoskite structure of type ABO_3 is known to be exhibited by ferroelectrics such as $BaTiO_3$, $PbTiO_3$, etc. [46]. Similarly, many pervoskites are known to exhibit magnetic ordering. However, both the orderings i.e., FE and magnetic are found to be mutually exclusive in pervoskites [47, 48]. This is because while the magnetic ordering requires incomplete d shell configuration, the FE ordering needs an empty d shell (such as Ti^{4+} in $BaTiO_3$) in transition metal ions. The ferroelectricity is on the account of the transition metal ion which is involved in covalent bond formation using its empty d shell orbitals with the oxygen. Consequently, the off-center displacement of the transition metal ion gives rise to ferroelectricity. Whereas, the existence of d electron will subdue the ferroelectricity mechanism up to some extent. The probable answer to the 'd^o V. d^n' problem underlies in identifying and synthesizing mixed pervoskites which would exhibit both d^0 and d^n transition metal ion configurations simultaneously. However, even in this case, the ME coupling will be weak.

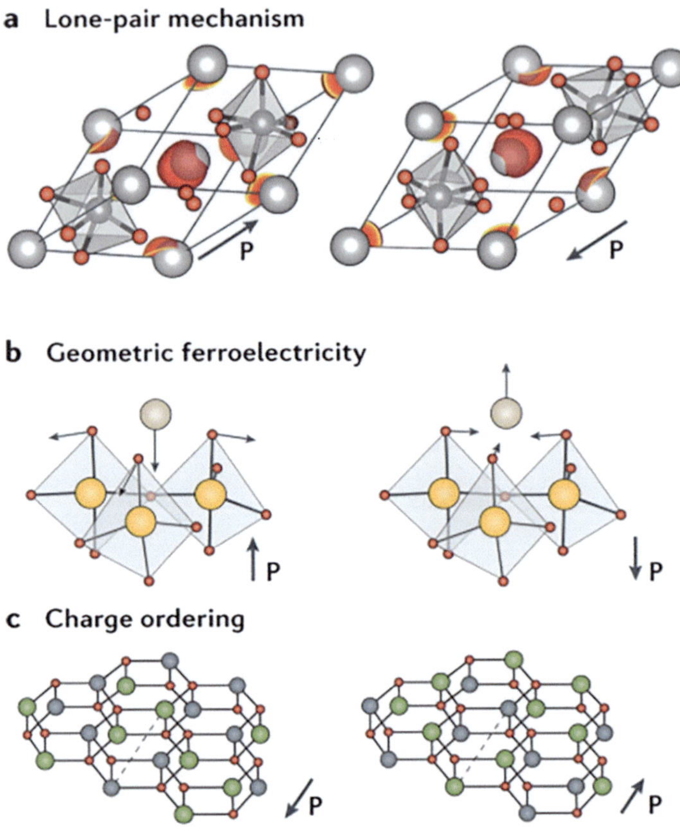

Fig. 1.11 Mechanism of the ferroelectricity in type 1 MFs. **a** Lone pair mechanism, **b** charge ordering, and **c** geometric ferroelectricity. Adapted with permission from Ref. [65]

(ii) **Lone pair mechanism**: The lone pairs create a local dipole moment resulting in a net spontaneous polarization such as the 'lone pairs' in Pb^{2+} in $PbTiO_3$ and Bi^{2+} ($6s^2$) in $BiFeO_3$. Herein, the lone pairs or dangling bonds, as they are often called, do not participate in chemical bond formation but rather form a local dipole. Thus, the dipole moment and consequently, high polarizability is obtained.

(iii) **Geometric ferroelectricity**: Here, the geometrical effects drive ferroelectricity. The examples include $BaNiF_4$ and h-$RMnO_3$ or hexagonal magnanites (R can be Y, Sc, In, Dy-Lu, Ho-Lu). In h-$RMnO_3$, the ferroelectricity is ascribed to unit cell tripling. Usually, h-$RMnO_3$ exhibits a FE behavior at high Curie temperature (T_C more than 1200 K) and low magnetic transition temperature (T_N less than 120 K) [49, 50]. One good example of the geometric ferroelectricity is $YMnO_3$, where the ferroelectricity occurs on account of MnO_5 polyhedra that gets buckled up, and not due to the Mn^{3+} ions (both Y^{3+} and Mn^{3+} do not possess lone pair) [51]. It comprises alternate layers of MnO_6

with an in-plane O^{2-} getting linked with apical O^{2-} ions and layer of Y^{3+} ions. The tilt of MnO_5 polyhedra occurs to form a close packed structure which results in O^{2-} ions moving near to Y^{3+} ions. The two Y–O bond lengths (with respect to in-plane O^{2-}) are different i.e., one is shorter and another larger, thus resulting in ferroelectricity. In addition, with A-type AFM ordering, $YMnO_3$ is thus regarded as a MF.

(iv) **Charge Ordering**: It occurs in the transition metal oxides with valence electrons, whose spatial arrangement results in an unequal distribution of charges. This results in the geometric frustration thereby, leading to ferroelectricity. $LuFe_2O_4$ (R-3m space group) is a good example which exhibit FE properties below 330 K [52, 53]. Its structure comprises of triangular lattices of ions i.e., Lu, Fe, O stacked in an alternative fashion. Both the Fe^{2+} and Fe^{3+} ions are arranged alongside the rhombohedral axis causing charge imbalance. This results in the charge frustration on the triangular lattice due to the geometrical configuration. The centers of the Fe^{2+} and Fe^{3+} ions do not concur and henceforth, we observe ferroelectricity. Also, the ferrimagnetic transition occurs below 240 K. Other examples include $Pr_{1-x}Ca_xMnO_3$ (mixed maganites), and $RNiO_3$ (Nickelates).

1.7.2 Type-II Multiferroics

In type-II MFs, the ferroelectricity and magnetic ordering occur jointly and exhibit magnetically or spin driven multiferroicity. In the magnetic ordered systems, the spins and charge interaction effectuate non-centrosymmetry to be shifted to the electric lattices thereby giving rise to polarization, and eventually to the FE response. Here, the extensively studied systems include $REMnO_3$ (RE = Dy, Tb), $REMn_2O_5$ (RE = Bi, Y), $Ni_3V_2O_6$, $FeVO_4$, and $CoCr_2O_4$. Type II MFs can be classified into two categories, based on the FE mechanism undertaken due to the (i) spiral and (ii) collinear magnetic structures, as follows:

(i) **Spiral type II MFs**: Here, the cycloidal spiral structure due to the magnetic or spin frustration is the cause of magnetic ordering and is accompanied with the generation of spontaneous polarization. $TbMnO_3$ (space group is Pbmn) is one of the type-II extensively studied MF [54, 55]. The Neel temperature of $TbMnO_3$ is $T_{N1} = 41$ K below which we observe a sinusoidal AFM behavior whereby, the spins point in a particular direction but exhibit periodic variation in magnetic moment (due to Mn^{3+} spins). Also, as shown in the Fig. 1.12, this spin density wave (SDW) shape which occurs below 41 K, has the direction given by the wave vector $\boldsymbol{Q}$ (1, 0.28, 0). [56]. At temperature further below $T_{N2} = 28$ K, we observe that the Mn^{3+} spins trace out a cycloid on account of spin rotation that is parallel to propagation vector $\boldsymbol{Q}$. Further, here we observe polarization, $\boldsymbol{P}$, which in case of cycloidal spiral, can be expressed as:

$$\boldsymbol{P} \sim \boldsymbol{r}_{ij} \times [\boldsymbol{S}_i \times \boldsymbol{S}_j] \sim [\boldsymbol{Q} \times \boldsymbol{e}], \tag{1.23}$$

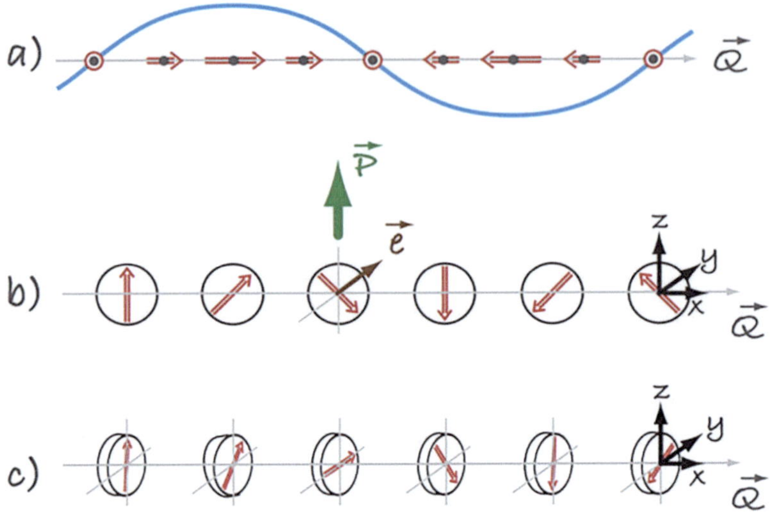

Fig. 1.12 Type II MF $TbMnO_3$ exhibiting, **a** sinusoidal SDW (below 42 K) is centrosymmetric and consequently, non-ferroelectric, **b** the cycloidal spiral structure (below 28 K) with spin-rotation in *x*–*y* plane and we observe polarization along *z* axis, **c** Herein, spin plane is perpendicular to the propagation vector and associated with inversion symmetry breaking. Adapted with permission from Ref. [42]

where, $\boldsymbol{S}_i$ and $\boldsymbol{S}_j$ are the two neighboring spin defined by a distance vector $\boldsymbol{r}_{ij}$, $\boldsymbol{e}$ defines axis of spin rotation, and $\boldsymbol{Q}$ is the propagation vector. We can infer from the Eq. 1.23 that the polarization vector lies in a plane of spins but perpendicular to the propagation vector as illustrated in Fig. 1.12. Thus, the variation in the applied magnetic field can influence electric polarization. Further, other examples besides $TbMnO_3$ of MFs with spiral magnetic structure include $TbMn_2O_5$, $MnWO_4$, $Ni_3V_2O_6$, $DyMnO_3$, etc. [57-59].

(ii) **Collinear type II MFs**: The collinear magnetic structures are also a source of ferroelectricity through the 'exchange striction' which can effectively break-up the magnetic ordering. One such example is Ca_3CoMnO_6 whose structure comprises of an alternative arrangement of 1D chain of Co^{2+} and Mn^{4+} ions [60, 61]. This geometry can be described by employing Ising model comprising of spin chains with nearest neighbor interaction being FM and further, the next nearest neighbor interactions being described as AFM. Consequently, the Ising spin chains exhibit $\uparrow\uparrow\downarrow\downarrow$ i.e., up-up-down-down type of magnetic ordering. Subsequently, this magnetic ordering can be broken-up by the exchange striction. The magnetic metal ions (Co^{2+}, Mn^{4+}) while alternating in between the spin chain facilitate superexchange interactions intermediated by O^{2-} ion. Here, the bond length is found to be shorter with respect to parallel spins and vice versa thus resulting in a net electric polarization and ferroelectricity.

1.8 ME Coupling

ME coupling refers to the co-existence of the magnetic and FE orders whereby the application of magnetic field can generate FE response in a material and vice versa [62, 63]. The mathematical analysis of ME effect can be effectively understood by taking into account tensor expansion of free energy of a homogenous medium with no underlying stress. The ME interaction in the system is expressed by the Gibbs free energy (ME interaction energy) equation [64]: The existence of ME coupling in single phase MFs such as: Cr_2O_3, $BiFeO_3$, $YMnO_3$, $BiMnO_3$, $RMnO_3$, RMn_2O_5, where 'R' is a rare earth element, is weak. For instance, Cr_2O_3 exhibits a weak ME coupling coefficient of 20 mV cm^{-1} Oe^{-1}. This has led researchers to fabricate composite MFs, in order to obtain a significant ME coupling with the potential to be utilized in ME-based device applications. Amongst the linear magneto-electro-elastic order coupling. Laminated ME structures have been fabricated. Giant ME effect is found to be exhibited by a laminate device fabricated from Pb(Zr, Ti)O_3-bimorph/NdFeB. The current ME laminate structures exhibit giant ME effect whose origin is magnetoelasticelectric. The applied magnetic field causes deformation of the magnetostrictive layer of the composite system. The deformation stress is then transmitted through elastic coupling to the piezoelectric layer. This results in an induced voltage across a dielectric layer. Beyond piezoelectric-ME ME laminates, we can utilize permanent magnets with a piezoelectric cantilever arrangement. The permanent magnets provide a static magnetic field which can interact with the externally applied ac magnetic field. As a consequence, we observe a magnetic force being induced into the piezoelectric cantilever arrangement resulting in a 'charge' output. The free energy expansion corresponding to a material is given by:

$$F\left(\vec{E}, \vec{H}\right) = F_o - P_i^s E_i - M_i^s H_i - \frac{1}{2}\varepsilon_o \varepsilon_{ij} E_i E_j - \frac{1}{2}\mu_o \mu_{ij} H_i H_j$$
$$- \alpha_{ij} E_i H_j - \frac{1}{2}\beta_{ijk} E_i H_j H_k - \frac{1}{2}\gamma_{ijk} H_i E_j E_k - \ldots \quad (1.24)$$

where, $\vec{E}$ and $\vec{H}$ are electric and magnetic field vectors, α_{ij} is a tensor form of linear ME coupling coefficient representing magnetic field induced polarization or electric field induced magnetization and β_{ijk} is the second order ME coupling coefficient. Also, ε_o represents the permittivity of free space, and ε_{ij} is rank second tensor of relative permittivity. Similarly, μ_o represents the permeability of free space, and μ_{ij}, is the rank second tensor of relative permeability. Interestingly, the third term describes the first order ME response whereby, α_{ij} is the linear ME tensor of rank two and bounded by relation

$$\alpha_{ij}^2 \leq \varepsilon_o \mu_o \varepsilon_{ii} \mu_{jj}. \quad (1.25)$$

Differentiation of Eq. (1.24) yields:

$$P_i\left(\vec{E},\vec{H}\right) = -\frac{\partial F}{\partial E_i} = P_i^s + \varepsilon_o \varepsilon_{ij} E_j + \alpha_{ij} H_j + \frac{1}{2}\beta_{ijk} H_j H_k + \gamma_{ijk} H_i E_j + \ldots \quad (1.26)$$

Similarly, we can obtain magnetization as follows:

$$M_i\left(\vec{E},\vec{H}\right) = -\frac{\partial F}{\partial H_i} = M_i^s + \mu_o \mu_{ij} H_j + \alpha_{ij} E_i + \beta_{ijk} E_i H_j + \frac{1}{2}\gamma_{ijk} E_j E_k \quad (1.27)$$

Here, P_i^s denote spontaneous polarization. M_i^s is spontaneous magnetization, ε and μ are electric susceptibility and magnetic susceptibility respectively.

Thus, ME coupling depends upon the variation in the grain size and shape. In addition, the strain induced piezoelectricity and magnetostriction as observed in the composite materials can also enhance the ME coupling coefficient value [65]. Subsequently, higher order ME coupling coefficient is represented by β_{ijk}, γ_{ijk}, and so on. Furthermore, often in multi-phase materials, we observe indirect coupling i.e., strain-mediated ME coupling order. Thus, the observation for the linear ME coupling coefficient involves measurement of magnetic response induced by electric or vice versa [66, 67]. Boomgaard and Born have identified some requirements for the two complementing phases showing an overall ME behavior, to get an efficient and superior ME response in ME-based **composite** systems [68]:

(i) The two required phases should possess a state of stable equilibrium with respect to similar sintering environment.
(ii) To keep the leakage current low, we require overall high resistivity of the composite system as in $BaTiO_3$–$CoFe_2O_4$ (BT-CFO) composites, BT has higher resistivity than the CFO phase.

Moreover, the crystallite size, buildup of space charges, interaction parameters between the two layers also influence ME coupling coefficient. In general, the magnetic response of a material can be controlled using the electric field in two ways (i) altering the spin canting (direction/amount), and (ii) modifying or regulating the exchange bias effect. In the materials exhibiting antiferromagnetism such as the $BiFeO_3$, we usually find a canted structure. However, in $BiFeO_3$, spin canting finds origin in Fe–O–Fe (FeO_6 octahedral) instead of FE origin. Consequently, the magnetic state cannot be influenced by following 180° switching of domains and thereupon the FE polarization. Figure 1.13 illustrates a general trend in magnetization and polarization values corresponding to type-I and type-II MFs including composite MFs.

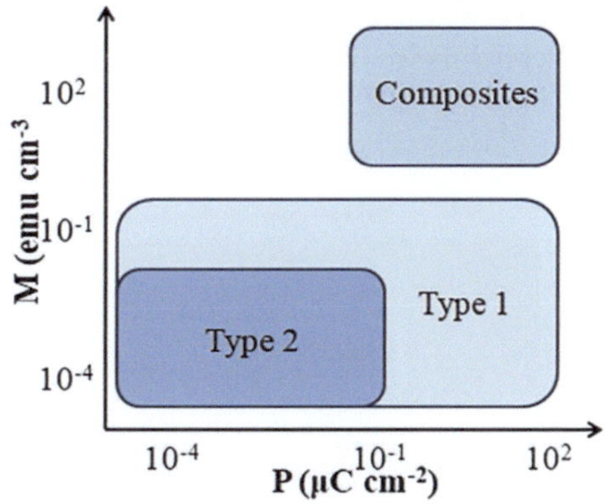

Fig. 1.13 A general trend as observed in magnetization and polarization values for the single-phase type I and type II MFs, and composite MFs

1.9 Some Basic Structures of MF and ME Materials

1.9.1 Single Phase MF, Bismuth Ferrite ($BiFeO_3$)

$BiFeO_3$ (BFO) is an extensively studied single phase MF. BFO possesses R3c space group and is a rhombohedrally distorted perovskite with FE ordering ($T_C = 1123$ K) and G-type AFM order along $[000]_h/[111]_c$ ($T_N = 643$ K). The hexagonal lattice parameters corresponding to the bulk BFO are $a = 5.5799$ Å and $c = 13.8670$ Å [40, 69, 70]. Origin of the ferroelectricity is through the lone pair mechanism whereby the lone pairs in the s orbital in Bi^{3+} form a local dipole. Thus, Bi^{3+} ions shift thereby resulting in net FE polarization along [111] direction. Figure 1.14 shows a rhombohedral unit cell whereby the two perovskite lattices are positioned along [111] direction. Also, seen is the rotation of oxygen octahedral in clockwise and anticlockwise by magnitude 13.8° and displacement of Fe by 0.135 Å along [111] direction. In addition, the G type of AFM ordering is due to the 'Fe' ions along with a canting of Fe–O–Fe angle. Even this small local magnetization is cancelled or averaged out to be zero due to the cycloidal spin structure. This cycloidal structure exists with the period of 620 Å in $\left[1\bar{1}0\right]$ direction for type 1 cycloid [71]. This inhibits the linear ME effect and only a weak quadratic ME coupling is observed [72, 73]. In general, the tolerance factor (t) for ABO_3 perovskites is given by the following equation:

$$t = \frac{r_A + r_O}{\sqrt{2}(r_B + r_O)}; \qquad (1.28)$$

where, r_A, r_B and r_O are the ionic radii with respect to the ions A, B, and C respectively of a perovskite structure of type ABO_3. Also, for ideal perovskites (cubic), the tolerance factor is 1.

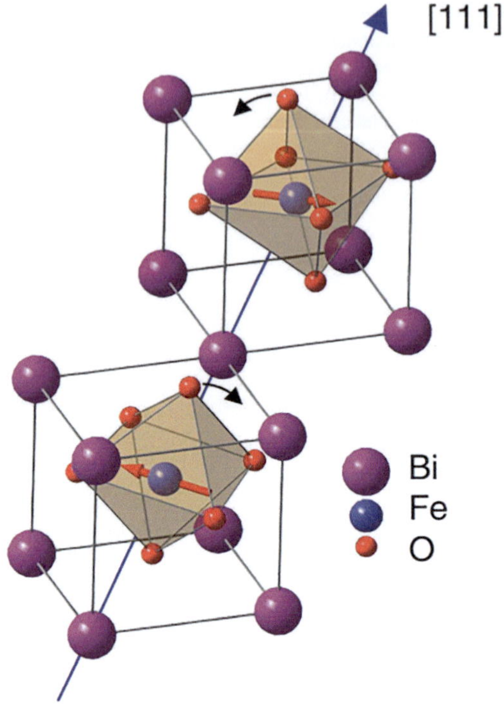

Fig. 1.14 Structure of single phase MF $BiFeO_3$. Adapted with permission from Ref. [83]

Generally, the 'A' site dopants in BFO include rare earths (such as La, Nd, Gd, Pr, Eu, Dy, Er, Tb, Sm, Ho) and IIA group metals (such as Ca, Pb, Sr, Ba) [74-78]. Here, the A-site doping such as Ca^{2+} which is non-magnetic, generates charge imbalance which can be compensated either by oxygen vacancies or by change in Fe ionic state resulting in distorted structure with reduction in spin canting of Fe–O–Fe bond angle and improved magnetic response [79]. The magnetic response in BFO is ascribed due to unpaired d electrons of Fe and associated with Fe–O–Fe superexchange interactions. The 'B' site dopants in BFO include Cr, Co, Mn, Zr, etc. [80-82].

1.9.2 Cobalt Ferrite ($CoFe_2O_4$)

$CoFe_2O_4$ (CFO) belongs to a wider group of spinel ferrites having general formula AB_2O_4 where the A site divalent cation could be occupying tetrahedral interstitial sites and B site trivalent cation could be occupying octahedral interstitial sites [84, 85]. CFO possesses cubic symmetry and an inverse spinel structure with O^{2-} ions arranged in an fcc lattice and space group Fd3m as shown in the Fig. 1.15. The Fe^{3+} cation exist both at the tetrahedral and the octahedral lattice sites and Co^{2+} cation at the ocatahedral site. CFO is a lucrative material on account of its interesting properties

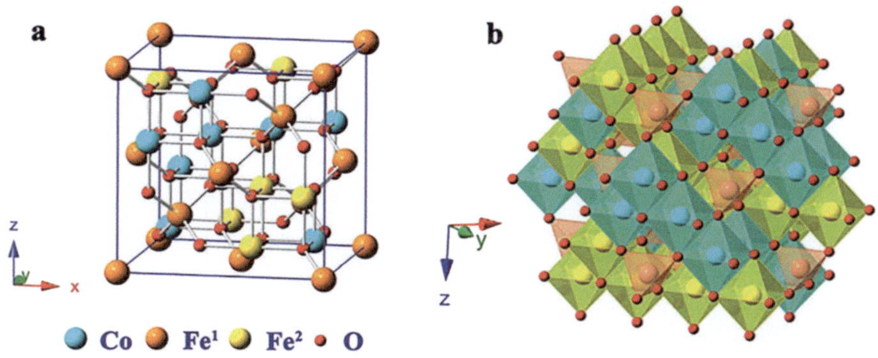

Fig. 1.15 Structure of $CoFe_2O_4$. **a** 'Fe^1' ions occupying 'A' sites, 'Fe^2' ions occupying 'B' sites. **b** O^{2-} ions arranged in an fcc lattice along with occupation of interstitial position. Adapted with permission from Ref. [86]

such as high coercivity, strong anisotropy, moderate saturation magnetization (M_S), chemically stable. The magnetic phase of CFO is generally included in composite systems in order to increase the overall magnetic properties such as BFO-CFO, BT-CFO, etc.

1.9.3 Barium Titanate, $BaTiO_3$

$BaTiO_3$ (BTO) has perovskite structure with space group P4mm. BTO exhibits polymorphs i.e., it shows rhombohedral symmetry up to 190 K followed by orthorhombic up to 278 K, and further tetragonal up to 395 K. Thus, at the room temperature, we observe only tetragonal polymorph with lattice constants $a = b = 0.3992$ nm and $c = 0.4036$ nm [87]. The Ti^{4+} cations occupy octahedral sites with Ba^{2+} cation exist in a twelve-coordination state as shown in the Fig. 1.16. The spontaneous FE polarization arises due to the displaced Ti^{4+} cation whose 3d orbital undergo hybridization with neighboring 2p orbitals of O^{2-} anions. Thus, though a good FE due to the bonding of empty 3d orbital of Ti^{4+} and filled p orbital of O^{2-} ($2p^6$), the empty d orbital does rule out the magnetic order.

1.10 Recent Advancement of Technologies, Applications and Device Fabrication

As discussed in the forthcoming chapters, MF and ME materials have been in exponentially rising demand owing to their multifarious applications. ME coupling is effectuated using three mechanisms namely, strain, charge carrier and spin exchange

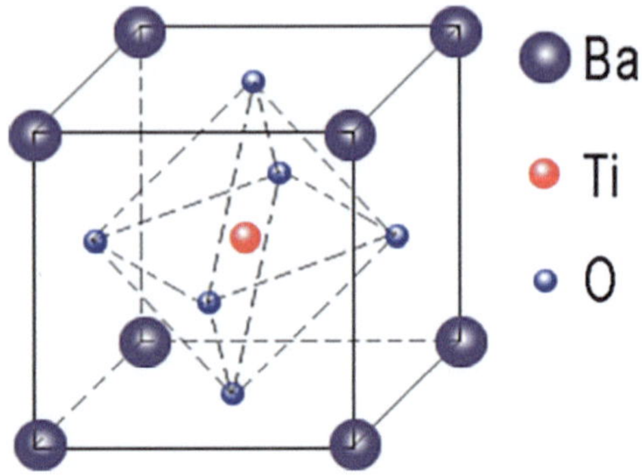

Fig. 1.16 Crystal structure of $BaTiO_3$. Adapted with permission from Ref. [88]

mechanism. The strain-mediated ME coupling occurs either by direct ME effect whereby the applied magnetic field induces magnetic phase strain via magnetostriction which gets transferred to the piezoelectric phase and finally resulting in electric displacement. Examples include ME-based magnetic field and current sensors, gyrators, energy harvesting devices, magnetic nanoparticles detection. Similarly, converse ME coupling is facilitated by an applied electric field that induces mechanical strain in the FE layer via inverse piezoelectric effect which is transferred to the magnetic layer thereby resulting in magnetic domain reorientation and magnetic field. Examples include spintronics-based spin nano-memory devices, MERAM devices, voltage adjustable inductivities and filters, bandpass filters, phase shifters, mechanical antennas, resonators. MF and ME materials are the future alternative memory applications amongst numerous other applications as discussing in the subsequent chapters. They can replace FE random-access memory (FERAM) while overcoming its inherent disadvantage of volatile reading of polarization. Another interesting topic is multiphase MF and ME materials which can be tailored for specific technical requirements in spintronic devices. For instance, the magnetic states can be effectively controlled in the spintronic devices employing ME materials such as MERAM and MEXOR digital gates. Recently, a single-phase MF, hexagonal wurtzite MnSe i.e., w-MnSe is an altermagnet semiconductor exhibiting ferroelectricity and a direct bandgap of 1.98 eV [89]. It shows good thermal stability as indicated by its FE critical temperature of 965 K. This can be a way forward for MF devices operating at room temperature and above. Similarly, the AFM domains in $BiFeO_3$ can be controlled by an electric field which is a useful property to exploit room temperature and efficient energy usage in a spin-valve device [90, 91]. Two dimensional (2D) MFs are known for device applications especially for information storage including fabrication of logic gates. For instance, 2D type-I MF namely, NbPc-covalent organic framework (NbPc COF) has been found to exhibit AFM and above room temperature ferroelectricity with ME coupling of order of two to three greater than other type-I MFs [92]. Another interesting development is the evidence of MF behavior in MXenes such as $Hf_2VC_2F_2$ which is a type-II MF, density functional theory predicted MF i-MXene i.e., $(Ta_{2/3}Fe_{1/3})CO_2$ is type-I MF exhibiting antiferromagnetism and FE polarization of ~12.33 $\mu C/cm^2$, etc. [93, 94]. Reportedly, MF tunnel junctions

(MTJs) which consist of a FE barrier and FM electrodes have been successfully fabricated [95]. It is fabricated with a 2D MF $CuCrSe_2$ that is an A-type AFM and also exhibit FE anti-FE (AFE) properties in its ground state. It is observed that this MTJ shows TMR of 2873.38% and TER i.e., Tunneling Electroresistance of 3410.71%. This has an immense potential to be utilized in non-volatile logic memory devices as well. Further, in a quest to achieve ultra-low power consumption logic and memory devices, magnon, associated with a spin wave, and magnon-based computing has received significant attention amongst the researchers worldwide [96]. Here, a magnon-mediated spin torque (MMST) MF logic device was fabricated using MF $BiFeO_3$ and can be uitilised for memory computing applications. More recently, ME-based sensors have also been fabricated through laser cutting method and have been successfully utilized in the detection of magnetic field. Numerous other applications include current sensing, position and proximity sensing, industrial control, fabrication of light weight, small and power efficient ME-based tunable inductors, antennas, phase shifter, filters, wireless ME-based medical implants, energy harvesting and storage devices [97]. It is important to consider both the material and technological aspect while fabricating multifunctional devices. A significant number of ME devices are reportedly fabricated by combining the magnetostrictive phase like FeGaC, CoFeC with the new piezoelectric phase of AlScN. The subsequent chapters discuss all these effects and related applications and multifunctional devices in-depth.

1.11 Conclusions and Future Challenges

The unique properties of MF and ME materials have attracted immense research interest from the researchers worldwide. The multifarious multifunctional device applications have been a major area of interest with a potential to develop room temperature, cost effective, environment friendly, ultra-low power consuming devices. W-MnSe, a recently reported MF has a potential future generation multifunctional ME device owing to its excellent thermal stability. Recently, researchers have also reported a new and interesting class of material namely, FM polar metals. As the name suggests, these materials combine magnetic properties along with polar structural distortions thereby allowing to explore ME coupling and metallicity. The newly spurred research in 2D MF materials is still in the initial stages for applications in non-volatile logic devices and demands further exploration. Some examples of 2D MFs include, p-doped SnSe which exhibits ferrimagnetism and ferroelectricity, $FePX_3$ and $CuFeP_2X_6$ with 'X' = Se, S, Te; heterostructure $CrMoC_2S_6$ (AFM)/ Sc_2CO_2(FE); FeCl etc. They find potential applications in next-generation emerging spintronic and multifunctional ME devices. Many researchers have been exploring ultra-low power consumption and non-volatile memory devices capable of logic operations and in-memory computing based on magnon transport. However, there is still a need to develop methods to understand the magnon transport mechanism which can be utilized to manipulate the electrical excitation and magnon modes.

Further, there is still a need to discover similar ME and MF materials which can offer a good thermal stability, stronger room temperature ME coupling, integration of ME and MF materials at nanoscale in order to engineer next-generation nanoscale electronic devices. The subsequent chapters also provide indepth discussion on the challenges faced with MF and ME materials such as urgent need for development of room temperature MF and ME materials exhibiting strong ME coupling coefficient, novel MF order etc. This book comprehensively summarises the structural properties of MF and ME materials, types and mechanism of ME coupling, device applications, challenges and solutions and recent technological advancements.

References

1. Landau LD, Lifshitz EM (1959) Electrodynamics of continuous media, vol 8 (Course of theoretical physics), Fizmatgiz, Moscow
2. Dzyaloshinskii IE (1960) On the magneto-electrical effects in antiferromagnets. Sov Phys JETP 10:528–629
3. Astrov DN (1961) Magnetoelectric effect in chromium oxide 13:729–733
4. Schmid H (1994) Multi-ferroic magnetoelectrics. Ferroelectrics 162:317–338
5. Spaldin NA, Fiebig M (2005) The renaissance of magnetoelectric multiferroics. Science 309:391–392
6. Spaldin NA, Ramesh R (2019) Advances in magnetoelectric multiferroics 18:203–212
7. Hill NA (2000) Why are there so few ferroelectrics? J Phys Chem B 104:6694–6709
8. Kimura T, Goto T, Shintani H, Ishizaks K, Arima T, Tokura Y (2003) Magnetic control of FEpolarization. Nature 426:55–58
9. Hur N, Park S, Sharma PA, Ahn JS, Guha S, Cheong S-W (2004) Electric polarization reversal and memory in a multiferroic material induced by magnetic fields. Nature 429:392–395
10. Onuta TD, Wang Y, Long CJ, Takeuchi I (2011) Energy harvesting properties of all-thin-film multiferroic cantilevers. Appl Phys Lett 99:203506
11. Kozielski L, Bochenek D, Clemens F, Sebastian T (2023) Magnetoelectric composites: Engineering for tunable filters and energy harvesting applications. Appl. Sci, 13:8854. https://doi.org/10.3390/app13158854
12. Chu YH, Martin LW, Holcomb MB, Gajek M, Han SJ, Hu W, Zhan Q, Yang PL, Rodriguez AF, Scholl A, Wand SX, Ramesh R (2008) Electric-field control of a local ferromagnetism using a magnetoelectric multiferroic. Nat Mater 7:478–482
13. Chen Y, Gillette SM, Fitchorov T, Jiang L, Hao H, Li J, Gao X, Geiler A, Vittoria C, Harris VG (2011) Quasi-one-dimensional miniature multiferroic magnetic field sensor with high sensitivity a zero bias field. Appl Phys Lett 99:042505
14. Gillette SM, Fitchorov T, Obi O, Jiang L, Hao H, Wu S, Chen Y, Harris VG (2014) Effects of intrinsic magnetostriction on tube-topology magnetoelectric sensors with high magnetic field sensitivity. J Appl Phys 115:17C734
15. Xu B, Íñiguez J, Bellaiche L (2016) Designing lead free sntiferroelectrics for energy storage. Nat Comm 8:15682
16. Wang J, Neaton J, Zheng H, Nagarajan V, Ogale SB, Liu B, Viehland D, Vaithyanathan V, Schlom DG, Waghmare UV, Spaldin NA, Rabe KM, Wuttig M, Ramesh R (2003) Epitaxial $BiFeO_3$ multiferroic thin film heterostructures. Science 299:1719–1722
17. Shi D, Zhang Q, Ye YM, Luo L, Cheng EK-W, Lin D (2017) High-frequency current sensor based on lead-free multiferroic $BiFeO_3$–$BaTiO_3$ based ceramics. Measurement 104:287–293
18. Yang S-C, Kumar A, Petkov V, Priya S (2013) Room–temperature magnetoelectric coupling in single-phase $BaTiO_3$–$BiFeO_3$ system. J Appl Phys 113:144101

19. Kotnala RK, Gupta R, Chaudhary S (2015) Giant magnetoelectric coupling interaction in $BaTiO_3/BiFeO_3/BaTiO_3$ trilayer multiferroic heterostructures. Appl Phys Lett 107:082908
20. Huang YL, Nikonov D, Addiego C, Chopdekar RV, Prasad B, Zhang L, Chatterjee J, Liu HJ, Farhan A, Chu YH, Yang M, Ramesh M, Qiu ZQ, Huey BD, Lin CC, Gosavi T, Iniguez J, Bokor J, Pan X, Young I, Martin LW, Ramesh R (2020) Manipulating magnetoelectric energy landscape in multiferroics. Nat Comm 11:2836
21. Pan H, Ma J, Ma J, Zhang Q, Liu X, Guan B, Gu L, Zhang X, Zhang YJ, Li L, Shen Y, Lin YH, Nan CW (2018) Giant energy density and high efficiency achieved in bismuth ferrite-based film capacitors via domain engineering. Nat Comm 9:1813
22. Zvezdin AK, Logginov AS, Meshkov GA, Pyatakov AP (2007) Multiferroics: promising materials for microelectronics, Spintronics and sensor technique. Bull Russ Acad Sci: Phys 71:1561–1562
23. Valasek J (1921) Piezoelectric and allied phenomena in Rochelle salt. Phys Rev 17:475
24. Kurchatov IV (1933) Ferroelectrics. State Technical Theoretical Publishing House, Moscow
25. Busch G, Scherrer P (1935) A new rock salt electric substance. Naturwissenschaften 23:737
26. Scott JF (2007) Applications to modern ferroelectrics. Science 315:954–959
27. Bokov AA, Ye Z-G (2006) Recent progress in Relaxor ferroelectrics. J Mater Sci 41:31–52
28. Martin LW, Rappe AM (2016) Thin film ferroelectric materials and their applications. Nat Rev 2:16087
29. Uchino K, Nomura S (1983) Critical exponents of the dielectric constants in diffused-phase transition crystals. Ferroelectr Lett 44:55–61
30. Marcus PM, Moruzzi VL (1988) Stoner model of ferromagnetism and total-energy band theory. Phys Rev B 38:6949–6953
31. Meiklejohn WH (1962) Exchange anisotropy-a review. J Appl Phys 33:1328
32. Zener C, Heikes RR (1953) Exchange interactions. Rev Mod Phys 25:191
33. Zener C (1951) Interaction between the d-shells in the Transition metals. II. FM compounds of Manganese with Perovskite Structure. Phys Rev 82:403
34. Moriya T (1961) Anisotropic superexchange interaction and weak ferromagnetism. Phys Rev 120:91
35. Ederer C, Spaldin N (2006) Electric-field-switchable magnets: the case of $BaNiF_4$. Phys Rev B 74:020401(R)
36. Vaz CAF, Hoffman J, Ahn CH, Ramesh R (2010) Magnetoelectric coupling effects in multiferroic complex pxide composite structures. Adv Mat 22:2900–2918
37. Martin LW, Crane SP, Chu Y-H, Holcomb M B, Gaje M, Hujiben M, Yang C-H, Balke N, Ramesh R (2008) Multiferroics and magnetoelectrics: thin films and nanostructures. J Phys: Condens Matter 20:434220
38. Ramesh R, Spaldin NA (2007) Multiferroics: progress and prospect and prospects in thin films. Nat Mater 6:21–29
39. Nan C-W, Bichurin MI, Dong S, Viehland D, Srinivasan G (2008) Multiferroic magnetoelectric composites: historical perspective, status, and future directions. J Appl Phys 103:031101
40. Lawes G, Srinivasan G (2011) Introduction to magnetoelectric coupling and multiferroic films. J Phys D: Appl Phys 44:243001
41. Wang Y, Hu J, Lin Y, Nan C-W (2010) Multiferroic magnetoelectric composite nanostructures. NPG Asia Mater 2:61–68
42. Khomskii D (2009) Classifying multiferroics: mechanisms and effects. Physics 2:20
43. Martin LW, Ramesh R (2012) Multiferroic and magnetoelectric heterostructures. Acta Mater 60:2449–2470
44. Pratt DK, Lyn JW, Mais J, Chmaissem O, Brown DE, Kolesnik S, Dabrowski B (2014) Neutron scattering studies of the ferroelectric distortion and spin dynamics in the type-1 multiferroic perovskite $Sr_{0.56}Ba_{0.44}MnO_3$. Phys Rev B 90:140401(R)
45. Picozzi S, Yamauchi K, Sergienko IA, Sen C, Sanyal B, Dagotto E (2008) Microscopic mechanisms for improper ferroelectricity in multiferroic perovskites: a theoretical review. J Phys: Condens Matter 20:434208

46. Posada CMF, Castro A, Kiat JM, Porher F, Peña O, Alguero M, Amorin H (2016) A novel perovskite oxide chemically designed to show multiferroic phase boundary with room-temperature magnetoelectricity. 7:12772
47. Goodenough JB, Longo JM (1970) Magnetic and other properties of oxides and related compounds, Landolt-Bornstein, numerical data and functional relations in science and technology, New Series, vol 4. Springer, p 126
48. Mitsui T, Hellwege KH, Hellwege AM (1981) Ferroelectrics and related substances, Landolt-Bornstein, numerical data and functional relations in science and technology, New Series, vol 16. Springer, p 1
49. Lilienblum M, Lottermoser T, Manz S, Selbach SM, Cano A, Fiebig M (2015) Ferroelectricity in the multiferroic hexagonal manganites. Nat Phys 11:1070–1073
50. Fiebig M, Frohlich D, Kohn K, Leute S, Lottermoser T, Pavlov VV, Pisarev RV (2000) Determination of the magnetic symmetry of hexagonal maganites by second harmonic generation. Phys Rev Lett 84:5620–5623
51. Aken BBV, Palstra TTM, Filippetti A, Spaldin NA (2004) The origin of ferroelectricity in magnetoelectric $YMnO_3$. 3:164–170
52. Viana DSF, Gotardo RAM, Cotica LF, Santos IA, Olzon-Dionysio M, Souza SD, Garcia D, Eiras JA, Coelho AA (2011) Ferroic investigations in $LuFe_2O_4$ multiferroic ceramics. 110:034108
53. Patankar S, Choudhary RJ, Phase DM (2015) Resonant photoemission study of multiferroic $LuFe_2O_4$ across three-dimensional charge ordering. 106:172905
54. Malashevich A, Vanderbilt D (2008) First principles study of improper ferroelectricity in $TbMnO_3$. Phys Rev Lett 101:037210
55. Kimura T., Goto T., Shintani H., Ishizaka K., Arima T., and Tokura Y., Magneticcontrol of FEpolarization, Nature, 426, 55-58, 2003.
56. Senff D, Link P, Hradil K, Heiss A, Regnault LP, Sidis Y, Aliouane N, Argriou DN, Braden M (2007) Magnetic excitations in multiferroic $TbMnO_3$: evidence for a hybridized soft mode. Phys Rev Lett 98:137206
57. Chatterji T, Brown PJ (2014) Field dependence of the magnetic structure of $TbMn_2O_5$. J Appl Phys 116:203904
58. Taniguchi K, Abe A, Takenobu T, Iwasa Y, Arima T (2006) Ferroelectric polarization flop in a frustrated magnet $MnWO_4$ induced by a magnetic field. Phys Rev Lett 97:097203
59. Feyerherm R, Dudzik E, Aliouane N, Argyriou DN (2006) Commensurate Dy magnetic ordering associated with incommensurate lattice distortion in multiferroic DyMnO3. Phys Rev B 73:180401(R)
60. Choi YJ, Yi HT, Lee S, Huang Q, Kiryukhin V, Cheong S-W (2008) Ferroelectricity in an Ising chain magnet. Phys Rev Lett 100:047601
61. Kaushik SD, Rayaprol S, Saha J, Mohapatra N, Siruguri V, Babu PD, Patnaik S, Sampathkumaran EV (2010) Magnetoelectric coupling in Ca_3CoMnO_6. J Appl Phys 108:084106
62. Bichurin MI, Filipov DA, Petrov VM, Laletsin VM, Paddubnaya N, Srinivasan G (2003) Resonance magnetoelectric effects in layered magnetostrictive-piezoelectric composites. Phys Rev B 68:132408
63. Ryu J, Vasquez AC, Uchino K, Kim HE (2001) Magnetoelectric properties in piezoelectric and magnetostrictive laminate composites. Jpn J Appl Phys 40:4948
64. Eerenstein W, Mathur ND, Scott JF (2006) Multiferroic and magnetoelectric materials. Nature 442:759–765
65. Fiebig M, Lottermoser T, Meter D, Trassin M (2016) The evolution of multiferroics. Naat Rev 1:16046
66. Michel C, Moreau J-M, Achenbach GD, Gerson R, James WJ (1969) The atomic structure of $BiFeO_3$. Solid State Commun 7:701–704
67. Cheong S-W, Mostovoy M (2007) Multiferroics: a magnetic twist for ferroelectricity. Nature 8:13
68. Boomgaard JV, Born RAJ (1978) A sintered magnetoelectric composite material $BaTiO_3$–Ni(Co, Mn)Fe_2O_4. J Mater Sci 13:1538–1548

69. Bammannavar BK, Naik LR, Chougule BK (2008) Studies on dielectric and magnetic properties of $(x)Ni_{0.2}Co_{0.8}Fe_2O_4 + (1-_x)$ barium lead zirconatetitanate magnetoelectric composites. J Appl Phys 104:064123
70. Catalon G, Scott JF (2009) Adv Mater 21:2463–2485
71. Xu B, Dupe B, Xu C, Xiang H, Bellaiche L (2018) Revisiting spin cycloids in multiferroic $BiFeO_3$. Phys Rev B 98:184420
72. Li Y, Fang X, Cao M (2016) Sci Rep 6:24837
73. Li Y, Cao W-Q, Yuan J, Wang D, Cao M-S (2015) J Mater Chem C 3:9276
74. Nayek C, Tamilselvan A, Thirmal C, Murugval P, Balakumar S (2014) Origin of enhanced magnetization in rare earth doped multiferroic bismuth ferrite. J Appl Phys 115:073902
75. Cheng ZX, Li AH, Wang XL, Dou SX, Ozawa K, Kimura H, Zhang SJ, Shrout R (2008) Structure, ferroelectric properties, and magnetic properties of the La-doped bismuth ferrite. J Appl Phys 103:07E507
76. Chakrabarti K, Das K, Sarkar B, Ghosh S, De SK, Sinha G, Lahtinen J (2012) Enhanced magnetic and dielectric properties of Eu and Co co-doped $BiFeO_3$ nanoparticles. Appl Phys Lett 101:042401
77. Maurya D, Thota H, Garg A, Pandey B, Chand P, Verma HC (2009) Magnetic studies of multiferroic $Bi_{1-x}Sm_xFeO_3$ ceramics synthesized by mechanical activation assisted processes. J Phys: Condens Matter 21:026007
78. Uniyal P, Yadav KL (2009) Pr doped bismuth ferrite ceramics with enhanced multiferroic properties. J Phys: Condens Matter 21:405901
79. Sardar K, Hong J, Catalan G, Biswas PK, Lees MR, Walton RI, Scott JF, Redfern SAT (2012) Structural, spectroscopic, magnetic and electrical characterization of Ca-doped polycrystalline bismuth ferrite, $Bi_{1-x}Ca_xFeO_{3-x/2}$ $(x \leq 0.1)$. J Phys: Condens Matter 24:045905
80. Wei J, Haumont R, Jarrier R, Berhtef P, Dkhil B (2010) Nonmagnetic Fe-site doping of $BiFeO_3$ multiferroic ceramics. Appl Phys Lett 96:102509
81. Azough F, Freer R, Thrall M, Cernik R, Tuna F, Collison D (2010) Microstructure and properties of Co-, Ni-, Nb- and W-modified multiferroic $BiFeO_3$ ceramics. J Eur Ceram Soc 30:727–736
82. Mukherjee S, Gupta R, Garg A, Bansal V, Bhargava S (2010) Influence of Zr doping on the structure and FEproperties of $BiFeO_3$ thin films. J Appl Phys 107:123535
83. Lubk A, Gemming S, Spaldin NA (2009) First principle study of FEdomain walls in multiferroic bismuth ferrite. Phys Rev B 80:104110
84. Antoniak CS, Schmitz D, Borisov P, de Groot FMF, Stienen S, Warland A, Krumme B, Feyerherm R, Dudzik E, Kleemann W, Wende H (2013) Electric in-plane polarization in multiferroic $CoFe_2O4/BaTiO_3$ nanocomposite tuned by magnetic fields. Nat Commun 4:2051
85. Pullar RC (2012) Hexagonal ferrites: a review of the synthesis, properties and applications of hexaferrite ceramics. Prog Mater Sci 57:1191–1334
86. Zeng X, Zhang J, Zhu S, Deng X, Ma H, Zhang J, Zhang Q, Li P, Xue D, Mellors NJ, Zhang X, Peng Y (2017) Direct observation of cation distributions of ideal inverse spinel $CoFe_2O_4$ nanofibres and correlated magnetic properties. Nanoscale 9:7493–7500
87. Buttner RH, Maslen EN (1992) Structural parameters and electron difference density in $BaTiO_3$. Acta Cryst B48:764–769
88. Wang JJ, Meng FY, Ma XQ, Xu MX, Chen LQ (2010) Lattice, elastic, polarization, and electrostrictive properties of $BaTiO_3$ from first principles. J Appl Phys 108:034107
89. Bezzerga D, Khan I, Hong J (2015) High performance room temperature multiferroic properties of w-MnSe altermagnet. Adv Funct Mater 2505813
90. Heron JT, Bosse JL, He Q, Gao Y, Trassin M, Le L, Clarkson JD, Wang C, Liu J, Salahuddin S, Ralph DC, Schlom DG, Iniguez J, Huey BD, Ramesh R (2014) Deterministic switching of ferromagnetism at room temperature using an electric field. Nature 516:370–373
91. Yu P (2025) Revolutions of multiferroic materials. NPJ Spintron 3:9
92. Li W, Zhu D, Dong S, Zhang J-J (2025) Two-dimensional multiferroic NbPc covalent organic framework with strong magnetoelectric coupling and room-temperature ferroelectricity. Phys Rev B 111:035439

93. Zhang J-J, Lin L, Zhang Y, Wu M, Yakobson BI, Dong S (2018) Type-II multiferroic $HF_2VC_2F_2$ MXene monolayer with high transition temperature. J Am Chem Soc 140:9738–9773
94. Zhao M, Chen J, Wang S-S, An M, Dong S (2021) Multiferroic properties of oxygen-functionalized magnetic i-MXene. Phys Rev Mater 5:094408
95. He K, Liu X, Yu G, Zhang Y, Zhang Y, Zhu S, Shen L (2025) Fully electronically controlled high-efficiency multiferroic tunnel junctions composed of bilayer $CuCrSe_2$ with anti-ferromagnetic and anti-ferroelectric properties. ACS Appl Nano Mater 8:9145–9154
96. Chai Y, Liang Y, Xiao C, Wang Y, Li B, Jiang D, Pal P, Tang Y, Chen H, Zhang Y, Bai H, Xu T, Jiang W, Jiang W, Skowronski W, Zhang Q, Gu L, Ma J, Yu P, Tang J, Lin Y-H, Yi D, Ralph DC, Eom C-B, Wu H, Nan T (2024) Voltage control of multiferroic magnon torque for reconfigurable logic-in-memory. Nat Commun 15:5975
97. Wu J, Qiao J, Zou S, Liang X, Ge B, Chen J, Du Y, Xu Y, Wang Z, Guo J, Hu Z, Liu M (2025) Magnetoelectric sensors with hollowed-out resonant structure fabricated by laser cutting. IEEE Sens J 25:26542–26552

Chapter 2
Synthesis Techniques and Measurements for Various Physical Properties of Multiferroic and Magnetoelectric Materials

2.1 Introduction

As the quest for innovative materials with superior functional capabilities intensifies, the requirements for engineered materials are evolving. These materials must demonstrate a broad spectrum of stable and flexible characteristics in order to guarantee their effectiveness in a variety of applications [1]. The rise of artificial intelligence (AI) and machine learning (ML) has driven the search for smaller, energy-efficient materials to replace complex semiconductor technologies in electronics and data storage. Over the past more than fifteen years, MFs have stood out for their unique physical properties and technological potential attracting the scientific community [2]. Their remarkable capability to display multiple properties at once, including ferromagnetism or antiferromagnetism, ferroelectricity and ferroelasticity, sets them apart from other materials. This unique characteristic enables a diverse range of potential applications across various domains including photovoltaics (PVs), optics and electronic technologies [3]. Attaining MF characteristics within a single phase presents significant challenges and is sporadic, often leading to suboptimal functional performance. A more promising strategy is to integrate materials with distinct properties into composite structures which can take the form of thin layers, laminates or bulk configurations.

In case of composites, ME-MF composites are generally formed by integrating the piezoelectric attributes of FE materials with the magnetostrictive features of FM materials making them suitable for applications like actuators, sensors, microwave phase shifters, memory devices and communication filters. The MF response of the system arises from the variation in capacitance within the FE material which is influenced by strain induced by an external magnetic field, and conversely, the effect of the electric field on the magnetic properties. This reciprocal relationship between the electric and magnetic components is fundamental to the functionality of ME-MF composites making them a focal point for research and application in advanced materials science [1, 4–7].

M. Kumar and S. Dabas, *Multiferroic and Magnetoelectric Materials*, Materials Horizons: From Nature to Nanomaterials, https://doi.org/10.1007/978-981-95-4391-5_2

MFs are generally divided into two main types: single-phase and composite MFs. These classifications exhibit notable differences in symmetry, structural composition, design and various physical or chemical properties. Single-phase MFs feature the coexistence of ferroelectricity or ferroelasticity alongside magnetism within a singular phase. On the other hand, composite MFs are characterized by the integration of multiple phases, each possessing unique order parameters. This combination is essential for the manifestation of the ME effect which arises from the interaction between the different phases. The distinct properties of each phase contribute to the overall functionality of the composite material. The primary challenge with single-phase MFs lies in their typically weak ME coupling or the occurrence of ME coupling only at low temperatures. However, recent developments have introduced new single-phase MFs exhibiting ME coupling at ambient temperatures. Additionally, composite MFs, particularly nano- and heterostructure demonstrate stronger ME coupling at ambient temperatures through cross-coupling interactions between magnetic and FE phases with a focus on thin films for practical applications [8]. According to Khomskii's classification, MFs are categorized into two types: type-I and type-II as discussed in the previous chapter [9–21] and also structural wise the classification is based on single perovskites, double perovskite and 2D materials-based perovskites.

2.1.1 *ABO_3 Type Perovskite MFs*

The most recognized FE materials are those belonging to the ABO_3 perovskite family including compounds such as $BaTiO_3$ (BT) and Pb(Zr, Ti)O_3 (PZT). Despite their prominence, these materials encounter significant challenges in practical applications [10]. $BiMnO_3$ is one of the notable MF materials but it faces significant challenges particularly the low temperature required for its MF properties to emerge [22]. The search for new MF compounds is increasingly focused on perovskite-based materials containing bismuth or lead. Typically, Bi^{3+} or Pb^{2+} ions occupy the A-site in the perovskite structure where their $6s^2$ lone pair electrons contribute to ferroelectricity [23, 24]. The $Bi_{0.5}Na_{0.5}TiO_3$ perovskite system despite exhibiting weak ferromagnetism has demonstrated enhanced FE properties following cation optimization which positions it as a viable candidate for various electronic applications. In contrast, materials like $BiCoO_3$ and $PbVO_3$, characterized by their tetragonal crystal structures and significant distortions may possess FE properties; however, the verification of their polarization changes remains elusive due to the presence of high coercive fields. In the realm of lead-free alternatives, the NiO–Mg magnetic nanocomposite stands out by exhibiting showcasing both FM and FE ordering at temperatures exceeding room temperature. This dual functionality enhances its potential for applications in advanced electronic devices. The exploration of such materials is crucial for developing sustainable technologies that do not rely on traditional lead-based compounds. Furthermore, the flexible nanocomposite PDMS-$BiMgFeCeO_6$ has been noted for its remarkable charge storage capabilities and MF properties [25–27].

2.1.2 Double Perovskite MFs

Double perovskite oxides characterized by the formula $A_2BB^{'}O_6$ exhibit a unique arrangement of cations B and $B^{'}$ that are organized in a rock-salt structure at the B site of the perovskite framework (ABO_3). This specific ordering contributes to the distinct properties of these materials and consequently, has significant interest in various applications. The unique structure of double perovskite oxides allows them to exhibit an array of intriguing physical phenomena [28]. Their semiconducting and magnetoresistive nature make them particularly interesting for applications in electronic devices while their ferromagnetic and half-metallic properties open avenues for advancements in spintronic and magnetic materials. Furthermore, the magneto-dielectric coupling observed in these compounds enhances their potential for multi-functional applications. As research continues to explore the capabilities of double perovskite oxides, their diverse properties position them as promising candidates for future technological innovations in various fields including electronics and materials science [29]. These materials are particularly intriguing because the B and B′ cations can be ordered at the octahedral sites with A cations occupying the voids formed by BO_6 and $B^{'}O_6$ octahedra. This ordering arises from significant differences in the ionic radii and formal charges of the B and $B^{'}$ cations [16, 30].

Numerous studies have been conducted on various lanthanum-based double perovskites such as La_2NiMnO_6 (LNMO), La_2CoMnO_6 (LCMO) and La_2FeMnO_6 (LFMO). These materials demonstrate a significant interaction between FE and FM properties although they operate effectively at temperatures lower than room temperature [31–33]. A comparable issue arises with yttrium-based double perovskites such as Y_2MnCrO_6 (YMCO) and Y_2NiMnO_6 (YNMO) which exhibit FM transition temperatures near 50 K [34, 35]. The synthesis of Sr-based double perovskite Sr_2FeMoO_6 (SFMO) through a sol-gel combustion method also encounters challenges as highlighted by Alizad et al. Their findings indicated the presence of a secondary phase, $SrMoO_4$, in addition to the primary Sr_2FeMoO_6 perovskite that possesses tetragonal symmetry. The emergence of this secondary phase complicates the synthesis process of Sr-based double perovskites which are characterized by only a limited interaction between their FM and FE properties. This limitation poses significant challenges for researchers aiming to optimize the performance of these materials in various applications. The interplay between the different phases can hinder the desired properties that are essential for effective utilization [36].

In the pursuit of room-temperature MFs, researchers have increasingly turned their attention to ordered double perovskite oxides that incorporate bismuth at the A-site. The unique characteristics of the $6s^2$ lone pair electrons play a crucial role in facilitating ferroelectricity. Among the various candidates in the realm of bismuth-based double perovskites, Bi_2FeCrO_6 has been specifically examined by Wu et al. [37]. Theoretical investigations have proposed that Bi_2FeNiO_6 may show the unique ability to function as both a FE and FM at room temperature [28, 38].

2.1.3 2D MFs

Dimensionality plays a crucial role in material properties especially in MFs. The discovery of 2D FM materials has been a major breakthrough, driving intense interest in 2D FE and MF materials. In the realm of traditional FEs and MFs, research is predominantly conducted in a three-dimensional setting where the importance of dimensionality tends to be overlooked. Conversely, the study of 2D van der Waals materials accentuates the FE instability allowing for a more nuanced understanding of its behavior in lower dimensions [39–41]. While research on 2D MFs is still in its early stages and largely theoretical, initial findings on ME coupling in 2D systems are varied. However, the concept of "magnetic reading and electrical writing" may be more feasible in 2D MFs [42]. Many 2D MFs are expected to be type-I where magnetism and ferroelectricity are independent. Ferroelectricity and consequently multiferroicity might be achieved through the covalent functionalization of nano-magnets. Broken inversion symmetry enables horizontally switchable polarization which formed the basis for the first 2D MF proposed in 2013 [43]. Examples of 2D MFs include CrN, CrB_2, $VOCl_2$ monolayer and CrI_3.

Recent forecasts have highlighted the presence of multiferroicity within the family of transition metal thiophosphates (TMTP). In particular, the monolayer $CuMP_2X_6$ where M represents either Cr or V and X denotes S or Se, exhibits characteristics of both ferroelectricity and ferromagnetism. As of now, the MXene $Hf_2VC_2F_2$ monolayer is recognized as one of merely two known examples of 2D type-II MFs [44]. Predictions also suggest similar ME coupling in metal-intercalated MoS_2 bilayers. Recent studies suggest that the net magnetization in 2D MF materials can be altered by employing FE switching techniques [45, 46]. A recent investigation suggests that a collection of van der Waals bilayers such as boron nitride (BN) and molybdenum disulfide (MoS_2) may serve as 2D FEs capable of generating an interlayer voltage. Furthermore, certain magnetic bilayers including MXenes, vanadium disulfide (VS_2), molybdenum nitride (MoN_2) and lanthanum chloride/bromide (LaCl/LaBr) have been found to display MF characteristics [47]. Overall, this chapter covers the various synthesis as well as the measurement techniques for the ME-MFs in detail in the following sections.

2.2 Synthesis Techniques

2.2.1 Solid State Reaction

The solid-state method is the most widely used technique for synthesizing polycrystalline solid materials from solid precursors. The final formation of nanomaterials involves sintering which is usually near 0.7–0.85 times the melting point of the nanomaterial. Sintering facilitates strong inter-particle bonds resulting in decrease in the surface energy. However, the increased grain growth may also hinder the nanostructure formation. A major benefit of the solid-state reaction technique is its ability to facilitate large-scale production [48]. The reactions proceed as follows:

$$\begin{aligned} &\text{Interface 1: } MO/MFe_2O_4 \\ &\text{Interface 2 : } Fe_2O_3/MFe_2O_4 \\ &\text{Overall reaction: } MO + Fe_2O_3 \rightarrow MFe_2O_4 \end{aligned} \quad (2.1)$$

The symbol 'M' is used to represent any divalent metal cation indicating the reaction occurs at two interfaces amid the grinding operation. Both the thin films and bulk ceramic powders of $BiFeO_3$ (BFO) and lithium-doped BFO were synthesized utilizing a standard solid-state technique which was facilitated by the ball milling process lasting for four hours. The precursors utilized in this process included high-purity oxides of bismuth, lithium, and iron. The material underwent a calcination process in air atmosphere at 760 °C for 1.5 h.

Following the calcination, pulsed laser epitaxy fabrication technique was employed to deposit nanocomposite thin films of BFO-$LaFeO_3$ onto a single-crystal $SrTiO_3$ (001) substrate. The ceramic target used for this deposition was 9Li-BFO which played a significant role in achieving the desired material properties. The ceramic target underwent ablation utilizing a krypton fluoride (KrF) excimer laser which operated at a wavelength of 248 nm. The process was conducted with a fluence of 1.5 J/cm^2 ensuring effective material removal. During the entire deposition process, the temperature of the substrate was consistently maintained at 660 °C while the oxygen partial pressure was set at 80 mTorr. Following the completion of the deposition, the samples underwent a cooling process which brought them down to room temperature. This cooling was conducted under an elevated oxygen pressure of 100 Torr ensuring that the samples were adequately protected during the transition [3, 49].

2.2.2 *Hydrothermal Route*

Hydrothermal synthesis serves as a practical technique for generating nanoparticles involving the interaction of precursor solutions in a controlled, heated chamber at increased pressure. This method is particularly advantageous for synthesizing various sophisticated ferrite nanomaterials [50]. BFO was synthesized with 40 mL of solution taken from the mixture of $Bi(NO_3)_3$ and $Fe(NO_3)_3 \cdot 9H_2O$ in molar ratio 1:1 obtained with KOH acting as a mineralizer mixed with deionized water as the reaction medium. The mixture was then kept for ultrasonification for 15 min, and then transferred to a sealed Teflon autoclave. The autoclave containing the mixture was placed in a microwave oven operating at 2.45 GHz and 300 W (adjustable power). Hydrothermal treatment was conducted at 180 °C with varying soaking times. The reactions were carried out for 5, 30 min and 1 h [51].

The research team led by Cristian successfully synthesized bismuth ferrite nanostructures by employing a unique hydrothermal approach. Their methodology included mixing ferric and bismuth nitrates with a sodium hydroxide solution at a concentration of 2 M, supplemented by 30 mL of water and carried out in an autoclave. The solution that had been prepared was transferred into a hydrothermal vessel lined with Teflon and subjected to heating in a muffle furnace at a temperature of 200 °C for duration of 12 h. This process facilitated the transformation of the solution into solid form. Following the heating process, the resultant powders were collected and set aside for subsequent characterization [51]. A solution was prepared by dissolving one millimole each of iron and bismuth nitrates in 20 mL of distilled water under continuous stirring, resulting in an opaque mixture containing white particulates. To obtain a transparent solution, diluted nitric acid was gradually added to the mixture, facilitating the dissolution of the suspended particles. The solution obtained was agitated thoroughly as 40 mL of 8.0 M KOH was introduced gradually, leading to the formation of a brown suspension. This mixture was then allowed to blend for an extra two hours before being placed into a 100 mL autoclave. Following this, the autoclave was subjected to heating in an oven reaching a temperature of 200 °C over a period of six hours with a controlled increase of 5 °C per minute. After multiple washes with distilled water, the dark brown $BiFeO_3$ (BFO) particles were dried at 80 °C for 24 h [52].

3D microflowers of bismuth ferrite were created through a standard hydrothermal synthesis. Dissolve 30.0 mmol of bismuth nitrate pentahydrate ($Bi(NO_3)_3{\cdot}5H_2O$) and ferric nitrate nonahydrate ($Fe(NO_3)_3{\cdot}9H_2O$) in 5.0 mL of acidified distilled water. Add aqueous ammonia dropwise to the mixture until the pH reaches 10, leading to the formation of an orange-yellow precipitate. The washed precipitate was transferred to a Teflon-lined autoclave, sealed and heated at 160 °C for 14 h. Once the hydrothermal process was completed, the autoclave underwent a cooling phase until it reached room temperature. The resulting material was collected, rinsed with distilled water and dried at 80 °C for 4 h [50].

2.2.3 Sol-Gel

The sol-gel process involves two fundamental reactions: hydrolysis and polymerization of metallic precursors in an appropriate solvent. The sol-gel synthesis technique employs a wet chemical methodology that starts with a precursor solution and the addition of chelating agents. This process culminates in the formation of a gel which serves as a foundational structure for further material development and applications. Metal nitrates and alkoxides are commonly used as precursors in the sol-gel process. These react to form metal-hydroxo or metal-oxo chains resulting in a gel-like system with both solid and liquid phases. The morphology of the gel ranges from discrete particles to extensive polymer networks, influenced by factors such as temperature, pH and metal ion concentration. Changes in pH notably affect the porosity of the gel and surface properties [53].

MF BFO nanoparticles, both pristine and cobalt-doped, were produced through the incorporation of 0.008 mol of bismuth nitrate pentahydrate into a blend of acetic acid and 17 mL of ethylene glycol. The resulting mixture was subjected to stirring for duration of 1.5 h at ambient temperature. A total of 0.008 mol of iron nitrate nanohydrate and cobalt nitrate hexahydrate were dissolved in acetic acid while maintaining continuous stirring for duration of 1.5 h. Following this initial mixing, the two solutions were combined and stirred for an additional 3 h to achieve a consistent and transparent precursor solution with a concentration of 0.2 M. The prepared precursor solution was subsequently subjected to a drying process at a temperature of 120 °C resulting in the formation of a dry gel. This gel was then calcined at 600 °C for a period of 2 h to facilitate the necessary chemical transformation [54, 55]. The samples were prepared using the sol-gel combustion method with a fuel mixture of 25 mL 2-methoxy ethanol and 25 mL ethylene glycol, stirred for 30 min. To achieve optimal sample quality, the fuel mixture was added drop-wise to acetic acid while stirring, until a pH of 1 was reached. It was determined that 26.6 mL of acetic acid combined with 50 mL of the fuel mixture yields the desired pH. The pH value is used as a standard for other samples. After mixing the fuel and acid for 30 min, $Bi(NO_3)_3 \cdot 0.5H_2O$ is added and stirred for another 30 min, then $Fe(NO_3)_3 \cdot 9H_2O$ is introduced [3]. The solution transitioned from colorless to brick red, then crimson and finally to blackish red. After stirring continuously at 70 °C for 1 h, the solution was heated for 2 h producing a fluffy gel with brown fumes. The gel was then dried in a hot air oven at 120 °C for 70 min. The dried powder was calcined at 600 °C for 3 h, then cooled and ground in an agate mortar.

2.2.4 MBE Technique

MBE is a high-precision technique used to deposit thin single-crystal layers in an ultra-high vacuum. It involves directing atomic or molecular beams generated by evaporating liquid precursors, melting solids or sublimating substances onto a substrate for GeSn films as shown in Fig. 2.1 [56]. The choice of the precursor and its vapor pressure influenced by temperature are crucial for controlling the deposition rate and ensuring high-quality single-crystal layers. This method is essential for advancing materials science and semiconductor technology. In MBE, each beam delivers a specific element to the substrate. Because the beams travel a long distance in the chamber without colliding, they remain separate and unaffected by each other until they reach the substrate. This ensures precise and controlled layering of different elements in the thin film [57, 58]. MBE includes in situ tools like Auger electron spectroscopy (AES) and reflection high-energy electron diffraction (RHEED) to monitor and characterize the layers as they are deposited. Essentially, it's a technique for growing nanocrystals one atomic layer at a time. MBE can vary based on the source used: gas-source MBE (gaseous sources), metal–organic MBE (metal–organic compounds), chemical beam epitaxy (chemical beams) or solid-source MBE (solid sources). It features a unidirectional flow of beams with minimal collisions,

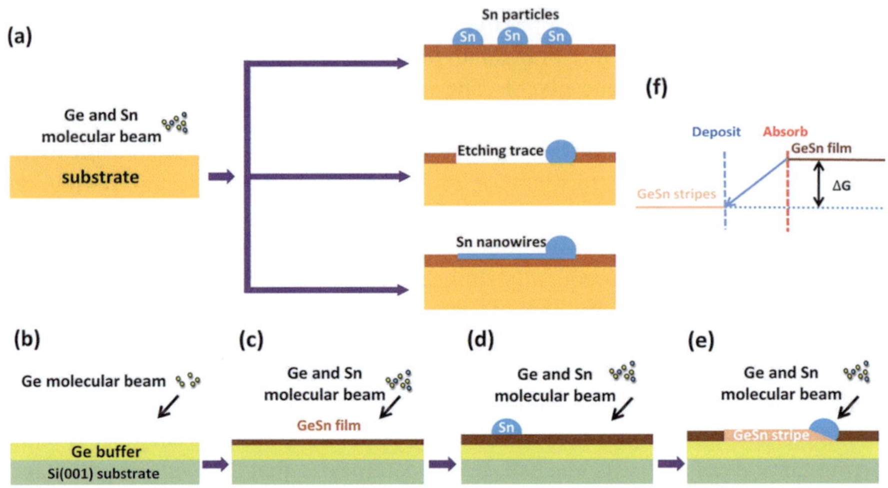

Fig. 2.1 **a** Schematic illustration of the Sn segregation phenomenon typically observed in low-Sn-content GeSn films during epitaxial growth. Due to the limited solubility of Sn in Ge, Sn atoms tend to segregate on the surface or form Sn-rich clusters, hindering uniform alloy formation. **b–e** shows the sequential schematic representation of the formation and evolution of a GeSn stripe during the epitaxial growth of a high-Sn-content $Ge_{0.8}Sn_{0.2}$ film. **f** Schematic diagram of the free energy difference between the uniform $Ge_{0.8}Sn_{0.2}$ film and the GeSn stripe structure. The stripe formation is energetically favorable, serving as the driving force for its spontaneous growth and structural evolution. Adapted with permission from Ref. [56] under the open access licence

as the process occurs in an ultra-high vacuum (typically below 10^{-10} Torr), allowing for extremely low impurity levels. The technique involves the reaction of beams with a heated substrate or FECA nanoparticle, requiring precise control of surface conditions to ensure high purity, sharp heterostructure, accurate composition control and uniform growth. Unlike CVD, growth of MBE rate is independent of substrate orientation and temperature but varies with diameter, typically ranging from 1 to 10 nm/min [59, 60]. Advancements in MBE have enabled the deposition of complex oxide thin films including doped and multicomponent oxides with multiple cations as well as bilayer and multilayer oxide films. While traditional high vacuum of MBE poses challenges for growing multicomponent oxides which often need higher oxygen pressures. This issue is mitigated by using highly oxidizing gases near the film surface while keeping the background pressure low [61, 62].

2.2.5 CVD Technique

CVD is a method utilized for the application of thin films onto various substrates through chemical reactions involving vaporized substances as illustrated in Fig. 2.2 [63]. The effectiveness of this technique is determined by several parameters such

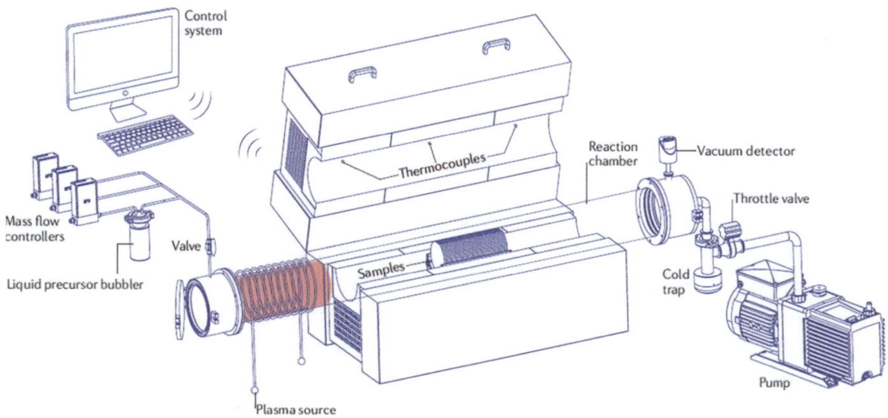

Fig. 2.2 Schematic diagram of a typical horizontal CVD system. The system consists of several integrated subsystems to enable precise thin film growth: gas delivery, quartz reaction chamber, vacuum, energy and auto-control systems. Adapted with permission from Ref. [63]

as the configuration of the reactor, the composition and ratios of gases as well as the pressure, temperature and duration of the growth process. CVD has gained significant traction in the production of few-layer and single-layer graphene films due to its efficiency and precision. Later, CVD was used to deposit a single layer of graphite on a platinum surface by decomposing hydrocarbons at 530 °C. Following the isolation of graphene in 2004, the potential of CVD for growing graphene films was demonstrated using LPCVD on iridium. The popularity of CVD for graphene growth is due to its ease of setup, successful long-term industrial use and scalability. Additionally, CVD is cost-effective and environmentally friendly for synthesizing graphene-based materials.

Now-a-days, various CVD methods are available, categorized into seven main types, based on factors like temperature, pressure, precursor type, gas flow, substrate temperature, deposition time, and activation source [64, 65]. A variety of precursors have been employed for the synthesis of graphene in CVD reactors, encompassing solid, liquid and gaseous carbon sources. Solid precursors can be directly placed into the reactor. For example, researchers have illustrated the formation of graphene films on copper substrates through the use of spin-coated PMMA combined with thermal sublimation in the presence of hydrogen and argon gases. In another approach, hexane was used as a source of liquid carbon. The hexane underwent a vaporization process which allowed it to be effectively introduced into the reactor via bubblers. This approach facilitated the precise management of the carbon source during the reaction. The concentration of hexane was regulated by passing an inert gas such as argon, through the liquid. Despite these methods, managing the carbon sources effectively can pose significant challenges during the growth process. As a result, gaseous carbon sources particularly methane (CH_4) have gained popularity in graphene synthesis. Methane is typically delivered into the reaction chamber through a specialized gas delivery system allowing for more precise control over the growth conditions and enhancing the overall efficiency of the graphene production process [66, 67].

CVD techniques are also classified according to the type of energy source utilized for the deposition of thin films. This classification is essential for understanding the various methods and their applications in material science and engineering. One prominent method is Hot-Wall CVD where the entire reaction chamber is uniformly heated by a furnace. This heating facilitates the necessary conditions for substrate annealing and the decomposition of the feedstock material ensuring effective film growth. Another approach is Cold-Wall CVD which involves directly heating the substrate to the required growth temperature while maintaining the reaction chamber at a lower temperature. Additionally, Plasma-Enhanced CVD (PECVD) employs high-frequency voltage to generate plasma in a low-pressure environment producing reactive species essential for the deposition of materials like graphene through inelastic collisions within the reactor [64].

2.2.6 PLD Technique

Among the diverse range of vapor deposition methods, PLD stands out as a highly effective technique, commonly employed for the fabrication of intricate oxide thin films. Its unique capabilities allow for precise control over the deposition process making it a preferred choice in various applications. In the process of PLD, as illustrated in Fig. 2.3 [68], a laser beam is aimed at a ceramic target that possesses the required chemical composition. This interaction generates plasma within a vacuum chamber which plays a crucial role in the subsequent material transfer. This plasma, created in a controlled vacuum environment facilitates the movement of the target material onto the substrate. When a high-energy pulsed laser beam possessing photon energies greater than 100 eV, interacts with a target, it causes the expulsion of particles. The outcome of this interaction is the creation of plasma that takes on a distinctive plume configuration. When laser pulses collide with a target, their energy is transformed into thermal, chemical and mechanical excitations. This conversion leads to various outcomes such as the evaporation of substances, the process of ablation, the excitation of particles, the formation of plasma and the exfoliation of material layers [69]. The characteristics of the laser, along with the optical, topological and thermodynamic attributes of the target material, play a crucial role in the ablation process of PLD. As the laser interacts with the target, it generates a plume that consists of various species including atoms, ions, molecules and clusters. This plume expands rapidly within the vacuum chamber, facilitating the transfer of mass to the substrate. PLD systems are designed for convenience, featuring a rotating holder that allows for straightforward replacement of targets which is particularly beneficial for the deposition of multicomponent films. During the deposition process, the introduction of a reactive gas such as oxygen can be employed to ensure the appropriate stoichiometry of oxide thin films. The partial pressure of oxygen plays a significant role in shaping the plume; increased pressure tends to decrease the mean free path of the ejected

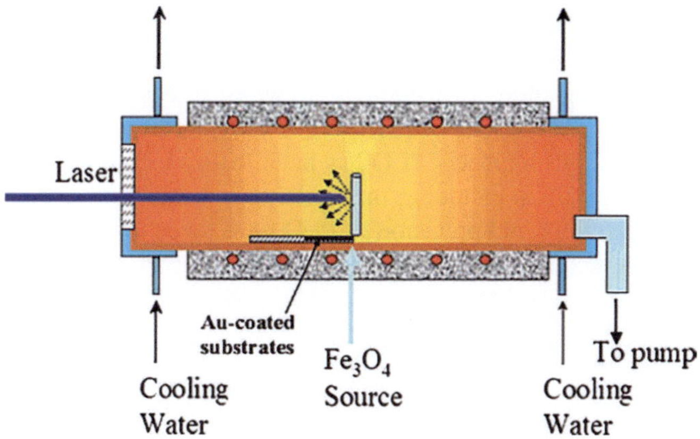

Fig. 2.3 Schematic illustrations of the PLD apparatus used for the synthesis of 1D nanostructure. In this setup, a high-energy laser beam passes through a transparent glass window and is focused onto a magnetite (Fe_3O_4) target inside a vacuum or controlled atmosphere chamber. Adapted with permission from Ref. [68]

particles resulting in a broader plume. To achieve the growth of textured or epitaxial films, it is often essential to heat the substrate during the deposition process. This heating can enhance the quality and structural properties of the films being deposited allowing for better control over the final characteristics of the material [70].

One of the primary benefits of PLD is its capacity to deposit a wide variety of materials as the non-equilibrium nature of the process effectively maintains the intricate stoichiometry of multicomponent systems. This technique stands out for its cost efficiency and relative simplicity when compared to other vapor deposition methods making it particularly suitable for the innovation of new materials and structures. Despite its advantages, PLD does have significant limitations particularly concerning the restricted area of uniformity which can complicate large-scale production efforts. This constraint can hinder the scalability of processes that rely on PLD for manufacturing as achieving consistent quality across larger substrates remains a challenge. Furthermore, if the erosion of the target material is not meticulously controlled, it can result in non-uniformities that lead to the formation of non-stoichiometric films. This issue underscores the importance of precise management during the deposition process to ensure the desired material properties are achieved consistently across the produced films [69, 70]. The integration of reflection high energy electron diffraction (RHEED) into the PLD process along with the incorporation of various emerging characterization techniques has significantly improved its functionality. This advancement allows for a more detailed analysis of the deposition process leading to better control over the material properties being developed. The application of PLD began in the 1960s, primarily focusing on the growth of semiconductor materials and basic perovskite oxide structures. This early utilization laid the groundwork for the technique but it was not until the late 1980s that PLD emerged as a significant method for thin film deposition across a range of structures.

During the late 1980s, PLD gained recognition for its effectiveness in producing high-quality thin films marking a pivotal moment in its development. This period saw an expansion in the applications of technique allowing researchers and engineers to explore its potential in various fields particularly in the fabrication of complex materials. Since its rise to prominence, PLD has been instrumental in the deposition of high-quality epitaxial films including multicomponent, MF bilayer and multilayer structures. The versatility and precision of PLD have made it a preferred choice for advancing material science and engineering enabling the creation of sophisticated thin film architectures [71–75].

2.2.7 *Floating Zone Technique*

A mixture of $BaCO_3$ (99.99% purity) and TiO_2 (99.99% purity) powders was prepared with a molar ratio of 1:2 for $BaCO_3$:TiO_2. This blend was subjected to ball milling in the presence of a small quantity of ethanol for duration of 260 ks. Following this process, the resulting mixture was transferred into a latex tube with a diameter of 10 mm and subsequently isostatically pressed to form a rod. The rod underwent a sintering process at a temperature of 1323 K for 36 ks in an air atmosphere. After sintering, the rod was melted and further solidified in a unidirectional manner utilizing the floating zone technique which employed a xenon arc lamp along with an ellipsoidal reflector. The rod was strategically placed at the focal point within a quartz tube chamber allowing for effective concentration of the optical radiation emitted by the xenon lamp onto the rod. To achieve a pronounced temperature gradient during the solidification process, a pair of aluminum foil was used to partially obstruct the heat radiation. This method is employed to control the thermal environment around the rod during the unidirectional solidification process. The careful manipulation of these parameters is crucial for optimizing the material properties of the final product. The radiation traversed the 10 mm space between the upper and lower masks, effectively heating the sintered rods. The growth rate recorded was 2.8×10^{-3} mm/s while the upper sintered rod was supplied at a rate a 3.6×10^{-3} mm/s. The upper and lower rods were set to rotate in opposite directions with the upper rod at 10 rpm and the lower rod at 3 rpm. Oxygen gas was introduced from the base of the quartz tube, moving upward at a flow rate of 2.5×10^{-5} m^3/s. Following the growth process, the rod underwent post-annealing in air at a temperature of 1073 K for a duration of 3.6 ks, after which it was sliced into a disc. The resulting disc was then subjected to analysis for phase and in-plane orientations using X-ray diffraction techniques. To achieve a smooth surface finish, the disc was mirror-polished using 1 μm diamond slurry resulting in a final thickness of 0.122 mm. This meticulous polishing process ensured that the disc was prepared adequately for further characterization and analysis. The combination of these processes highlights the precision involved in the preparation of the sintered rods and the subsequent disc [76]. The Laser Floating Zone (LFZ) technique facilitates the synthesis of materials that possess exceptionally high melting points by directing a CO_2 laser onto substances that effectively absorb this specific wavelength. This method guarantees the development of materials with accurate crystallographic orientation. Within a controlled growth chamber, ferrites are produced under varying atmospheric conditions and pressures which play a significant role in influencing the kinetics of crystallization. The growth chamber is equipped with mirrors that concentrate the laser light onto the sample ensuring optimal energy delivery for the melting process. A specialized software program manages the intensity of laser and the rate at which the material is pulled allowing for real-time monitoring of the molten area. This control is vital for achieving the desired crystal quality as the growth rate directly impacts the characteristics of the resulting crystals. A slower growth rate is generally associated with the formation of larger and more uniform crystals which tends to exhibit fewer secondary phases. However, the effectiveness of this approach is also contingent upon the stability of the phases of the material during the growth process. Thus, careful consideration of both the growth rate and the material properties is essential for optimizing the crystallization outcomes in the LFZ method [77]. This method is specifically developed to create strontium ferrite rods which serve as both seeds and feedstock for fiber development. The technique involves utilizing a laser beam to melt the precursor materials, thereby initiating the crystallization process. Throughout this operation, it is crucial to maintain precise control and monitoring of the growth rate to ensure optimal results. The composition of the strontium fiber is derived from iron oxide (Fe_2O_3, 99.8% purity) and strontium carbonate ($SrCO_3$, 99% purity) to achieve the desired stoichiometric ratio of $SrFe_2O_4$. After accurately weighing the components, they are combined in an agate vessel along with spheres and alcohol which acts as a dispersing agent. This mixture is processed in a planetary mill for two hours at a speed of 250 rpm to ensure a uniform powder blend. To enhance the binding of the particles, a solution of polyvinyl alcohol (PVA, 0.1 g/mL in distilled water) is incorporated into the mixture, resulting in a more cohesive formulation. This composite is then cold-extruded into rods measuring 3 mm in diameter and up to 12 cm in length exhibiting a texture similar to clay. Following a drying period at room temperature within a mold to maintain straightness, the rods are introduced into the LFZ system which is equipped with a 200 W CO_2 laser (Spectron SLC 200 W) operating at a wavelength of 10.6 μm. Fiber growth occurs at rates of 10, 50 and 100 mm/h under stable laser power, allowing for the evaluation of how these varying rates influence the material's electrical and magnetic characteristics. The feed rod rotates at 15 rpm while the seed rod rotates in the opposite direction at –5 rpm, resulting in fibers that display distinct color gradients: reddish at the base, dark grey during initial growth and metallic upon full formation [78].

2.2.8 Green Synthesis

The unique properties of Bi-based nanoparticles (BBNs) position them as highly advantageous in various technological fields. This has led to an increased focus on environmentally sustainable synthesis approaches for BBNs which are celebrated for their non-toxic and stable qualities as well as their ability to facilitate diverse production at the nanoscale [79, 80]. Yttrium-doped and undoped BFO nanoparticles of high purity were produced through an environmentally friendly approach utilizing a bio-template derived from seaweed specifically κ-carrageenan. The synthesized materials underwent thorough analysis and their effectiveness as photocatalysts was evaluated by testing their ability to degrade methylene blue (MB) under direct sunlight exposure. Characterization techniques such as Field Emission Scanning Electron Microscopy (FESEM) and High-Resolution Transmission Electron Microscopy (HRTEM) demonstrated that the nanoparticles exhibited a rhombohedral perovskite structure with dimensions ranging from 80 to 90 nm. These findings confirm the successful synthesis of the desired nanostructures using the green method. The photocatalytic efficiency of the synthesized nanoparticles was notably affected by the reduction in band gap energy which decreased from 2.07 to 2.05 eV as the concentration of yttrium dopant increased. This alteration in band gap energy suggests a potential enhancement in photocatalytic activity making these materials promising candidates for applications in environmental remediation [81]. Initially, guggul aqueous extract was converted into a Binary Graft Polymer (BGC), which served as the foundation for the subsequent synthesis. The next phase involved the integration of nano bismuth ferrite, leading to the formation of a Binary Interpenetrating Network (BIPN). This methodical approach facilitated the creation of the nanocomposite ensuring a structured development process. The enhanced dispersion of bismuth ferrite nanoparticles produced by BIPN when compared to BGC, contributes to its superior practical applications. Sourced from natural materials, this nanocomposite is not only highly biodegradable but also capable of decomposing after several uses making it an environmentally friendly option. In the process of creating the polymeric nanocomposite, 0.5 g of bismuth ferrite nanoparticles were incorporated during the synthesis of BIPN from BGC. The reaction was conducted under conditions that were specifically optimized for the polymers BGC and BIPN. Following the synthesis, the resulting product underwent a purification process which involved washing with double-distilled water to ensure the removal of any impurities. This step is crucial for achieving the desired quality and performance of the nanocomposite. The careful integration of bismuth ferrite nanoparticles into the polymer matrix not only enhances the properties of the final product but also demonstrates the effectiveness of the optimized reaction conditions. This approach highlights the importance of precise methodologies in the development of advanced materials [3, 82–84].

2.3 Measurement Techniques and Properties

2.3.1 Structural and Microstructural Properties

X-ray diffraction (XRD) techniques were employed to investigate the structural characteristics and crystalline phases of the single or multiphase materials. Kumar et al., discussed structural analysis through the XRD patterns of $(1-x)BiFeO_{3-x}BaTiO_3$ (BFO–BT) system at room temperature, as shown in Fig. 2.4a [24]. Notably, the sample with $x = 0.10$ exhibits clear peak splitting at specific 2θ angles, indicating a rhombohedral crystal structure. As the BT content increases, this peak splitting gradually diminishes. At $x = 0.30$ (BFO–30%BT), the emergence of a single sharp peak suggests a structural transition from the rhombohedral to the cubic phase.

Furthermore, a systematic shift of XRD peaks toward lower 2θ values is observed with increasing BT content, as depicted in Fig. 2.4b [24]. This shift is attributed to the substitution of the smaller Bi^{3+} ions (ionic radius = 1.03 Å) by larger Ba^{2+} ions (1.35 Å) at the perovskite A-site. In contrast, the Fe^{3+} (0.64 Å) and Ti^{4+} (0.605 Å) ions at the B-site have similar ionic radii, resulting in negligible influence on the peak position.

To minimize impurity phases, particularly $Bi_2Fe_4O_9$, its content was reduced to below 1% in BFO–(10–30)%BT samples by carefully controlling Bi_2O_3 volatilization during sintering. The synthesis involved calcination of $BaCO_3$ and TiO_2 at 820 °C for 8 h, followed by sintering at 940 °C for 2 h. A calcined powder of the

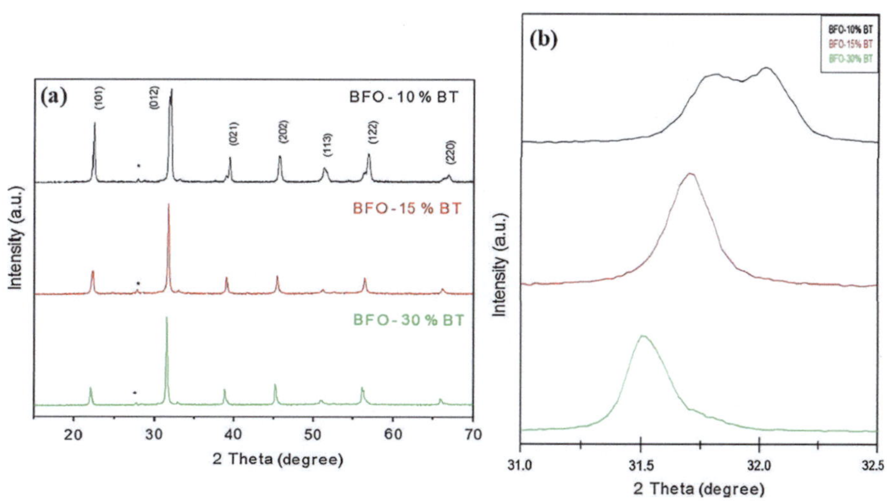

Fig. 2.4 **a** XRD patterns of BFO–BT solid solution samples with varying BT content: BFO–10%BT, BFO–15%BT, and BFO–30%BT, recorded at room temperature. The patterns indicate phase purity and allow assessment of structural changes induced by BT incorporation. **b** Magnified view of the (012) diffraction peak region, showing a progressive shift toward lower 2θ values with increasing BT content. Adapted with permission from Ref. [91]

same composition was used as a spacer to suppress Bi loss during the sintering process [24]. Further there are number of structural investigations are discussed via XRD in MF and ME based materials in the recent literature [1, 24, 85–90].

In a related study by Kumar et al. [85], the microstructural characteristics of the same BFO–BT solid solutions were examined using scanning electron microscopy (SEM). As shown in Fig. 2.5, the SEM images reveal a well-defined and homogeneous microstructure across all compositions. The average grain size was estimated to be approximately (3 ± 1) μm, indicating uniform grain distribution and effective sintering [85].

In the research carried by Dariusz Bochenek et al., the SEM micrographs of the MF composites are shown in Fig. 2.6 [1]. These images were captured using two techniques: standard SEM detection with secondary electrons (Fig. 2.6a–c) and

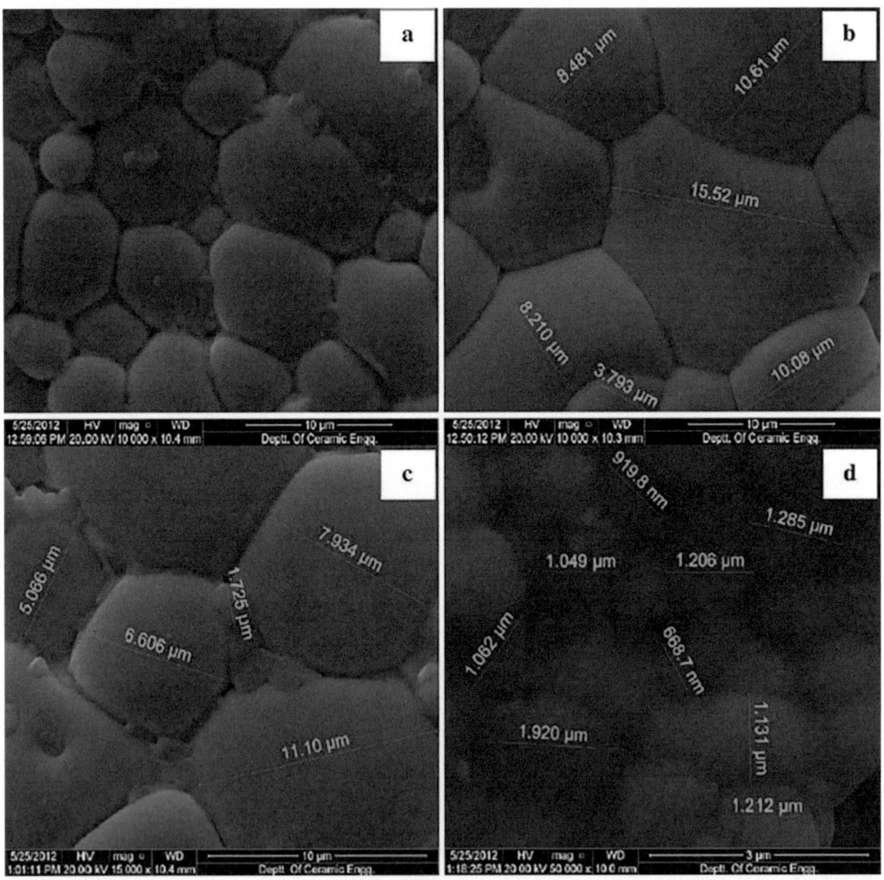

Fig. 2.5 SEM micrographs of BFO–BT solid solution samples with varying BT content, illustrating changes in surface morphology: **a** 0.9BFO–0.1BT, **b** 0.8BFO–0.2BT, **c** 0.75BFO–0.25BT, and **d** 0.7BFO–0.3BT. Adapted with permission from Ref. [85]

backscattered electron detection (BSE) (Fig. 2.6d–f). Typically, the BSE technique highlights areas in the microstructure with differing atomic numbers showing light and dark zones. However, in the case of the $BaTiO_3$–$Ni_{0.64}Zn_{0.36}Fe_2O_4$ (BT-F) MF composite, the BSE method is less effective, likely due to the minimal atomic number difference between BT and F. This contrasts with composites like PZT-ferrite where the magnetic phase is more distinct against the FE background. The ceramic composites exhibit strong sintering with grains joined together and no clear grain boundaries (Fig. 2.6a–c). Pores in the microstructure highlight limitations of high sintering temperatures. Figure 2.6g shows that as the ferrite content increases, the average ferrite grain size grows: 2.04 μm for 95BT-F, 2.26 μm for 90BT-F and 2.63 μm for 85BT-F. Larger ferrite content also leads to more agglomerates. The BT FE component shows greater grain size variation particularly for the 85BT-F composition with 15wt% ferrite (Fig. 2.6a, d) [1].

The investigation into the morphology and microstructure of the nanocomposites was also conducted through transmission electron microscopy (TEM). As per Dwivedi et al., in Fig. 2.7a [86], TEM image illustrates multiple nanoparticles of

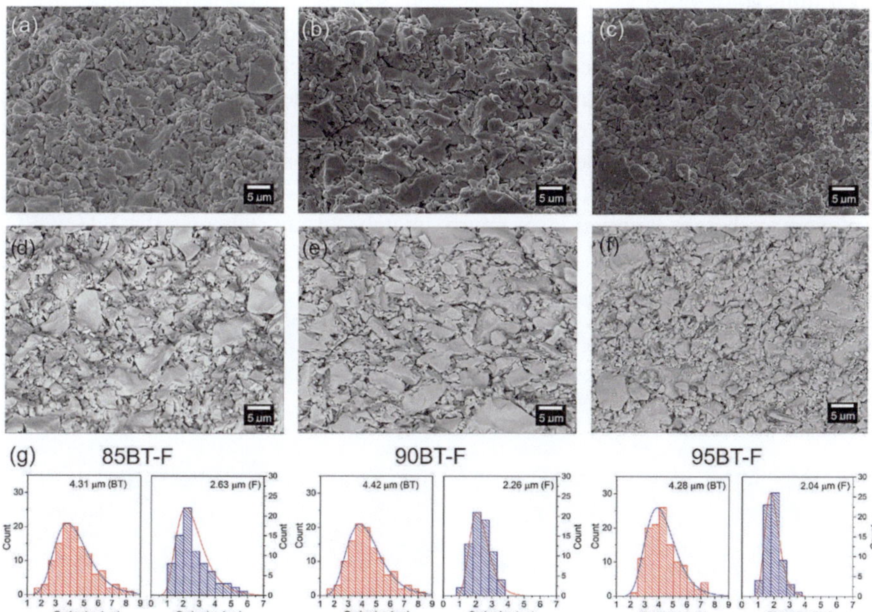

Fig. 2.6 SEM images showing the microstructure of BT–F MF composites with varying BT content: **a–c** Surface morphology observed using the **secondary electron (SE)** imaging technique for 85BT–F, 90BT–F, and 95BT–F compositions, respectively. **d–f** The corresponding backscattered electron (BSE) images for the same compositions, enhancing the compositional contrast between BT and the ferromagnetic (F) phase. **g** Grain size distribution analysis for the BT (red) and F (blue) phases in the BT–F composites, indicating distinct size populations and helping to evaluate the influence of BT content on microstructural development and phase dispersion. Adapted with permission from Ref. [1] under the open access licence

$La_{0.7}Sr_{0.3}MnO_3$ (LSMO) embedded within a 5% $PbZr_{0.52}Ti_{0.48}O_3$ (PZT) composite, referred to as (LP5). Upon detailed observation, it becomes evident that the majority of the nanoparticles exhibit a nearly spherical shape and possess smooth surfaces. The analysis presented in Fig. 2.7a illustrates the presence of nanoscale grains characterized by well-defined grain boundaries which may influence the magnetic characteristics of the material. The average particle size determined through TEM measurements, approximately 28 nm, aligns well with the estimates obtained from XRD analysis in this study. In Fig. 2.7b [86], a high-resolution transmission electron microscopy (HRTEM) image of an LP5 nanocomposite particle is displayed, revealing interplanar spacing of 0.277 and 0.414 nm. These measurements correspond to the (110) plane of $La_{0.7}Sr_{0.3}MnO_3$ and the (001) plane of $PbZr_{0.52}Ti_{0.48}O_3$, respectively. This observation substantiates the presence of two distinct phases within the composite material. The data suggests that the nanoparticles exhibit single crystalline characteristics and are predominantly devoid of significant lattice imperfection [86].

In a recent study conducted by Puli et al. [87], Fig. 2.8a, b illustrate the surface topography (z-height) images of BT and tri-layered BTO/$CoFe_2O_4$/BTO (BTO/CFO/BTO) films, respectively. These images were acquired using a multimode atomic force microscope (AFM) operated in contact mode, focusing specifically on the central regions of the ferroelectric (FE) and magnetoelectric specimens. The surface scans, each covering an area of $30 \times 30\ \mu m^2$, were recorded at a scan rate of 1 Hz with 512 samples per raster line, ensuring high-resolution and accurate characterization of topographical features. The AFM analysis revealed consistent grain growth across both film types, with microstructures free from pinholes and cracks. The BTO film displayed a z-height fluctuation of approximately 32 nm, whereas the BTO/CFO/

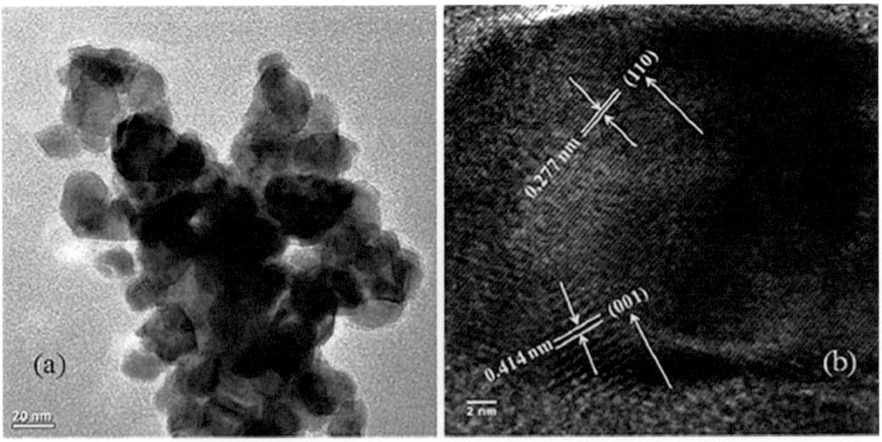

Fig. 2.7 **a** TEM image of the 0.95LSMO–0.05PZT nanocomposite, showing the overall morphology and nanoscale distribution of the LSMO and PZT phases. **b** HRTEM image of the same sample, revealing well-defined lattice fringes and clear interfaces between the LSMO and PZT phases. Adapted with permission from Ref. [86]

BTO film exhibited a more substantial fluctuation of 62 nm. Correspondingly, the root mean square (RMS) roughness values were measured at 16 nm for the BTO film and 21 nm for the BTO/CFO/BTO film, indicating increased surface roughness in the tri-layered structure. Both film types exhibited a relatively consistent average particle size, ranging from 60 to 180 nm, suggesting a uniform grain size distribution despite differences in surface morphology. The elevated surface roughness observed in the ME BTO/CFO/BTO film is likely attributed to its multilayer architecture and increased thickness. The contrasting crystal structures of BTO (tetragonal) and CFO (spinel), combined with the effects of varied annealing temperatures and pressures, significantly influence surface morphology by promoting diverse grain growth and microstructural features. Annealing conditions play a critical role in this context, with temperature and pressure impacting grain size evolution and surface irregularities. These processing parameters contribute to the overall roughness and complexity of the tri-layered film's surface, underscoring their importance in tailoring material properties for functional applications.

AFM imaging also revealed that the grain patterns in both films were randomly oriented with irregular and non-linear domains. Notably, the ME films exhibited larger domain sizes than their FE counterparts. This difference is likely due to the inherently smaller dimensions of ferroelectric domains compared to those in ferro- or ferrimagnetic materials. The larger domains observed in ME films may enhance ME coupling at domain boundaries, suggesting more complex interactions between FE and magnetic properties, and potentially improving the performance of these materials in ME-related applications.

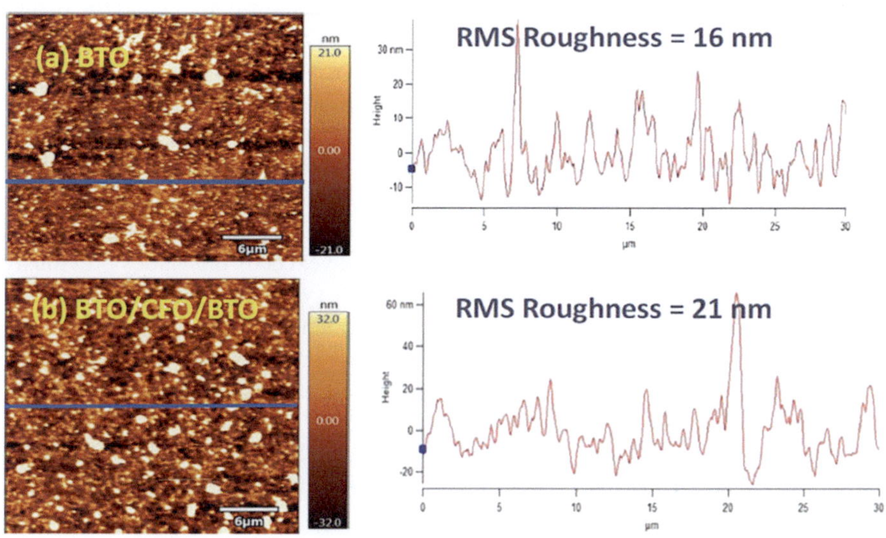

Fig. 2.8 **a** AFM z-height image showing the surface topography of a BTO thin film. **b** AFM z-height image of a BTO/CFO/BTO multilayer film, illustrating the influence of the CFO interlayer on surface morphology. Adapted with permission from Ref. [87]

In summary, the BTO/CFO/BTO films demonstrate increased surface roughness and domain size compared to BTO films, attributed to their multilayer structure, differing crystallography, and processing conditions. These morphological characteristics highlight the significant impact of material design and fabrication parameters on optimizing surface properties and functional performance [87–89].

2.3.2 *Magnetic and Ferroelectric Properties*

The magnetic and ferroelectric properties form the foundation of ME and MF material research. The coexistence of both ferroic orders—ferromagnetism and ferroelectricity—define a material as multiferroic, while the coupling between them gives rise to magnetoelectric behavior. In studies involving BFO–BT ME–MF solid solutions, Kumar et al. [85] investigated room-temperature magnetic behavior using a vibrating sample magnetometer (VSM) with an applied magnetic field of up to 5000 Oe. The resulting magnetic hysteresis loops are presented in Fig. 2.9. The analysis revealed that the saturation magnetization (Ms) decreases with increasing (BT) content (x) in the BFO–BT solid solution. This reduction in Ms can be attributed to multiple factors: distortion of the rhombohedral crystal structure, random distribution of $Fe3^{+}$ and $Ti4^{+}$ ions at the octahedral sites, and the presence of canted antiferromagnetic (AFM) ordering within the Fe–O–Fe spin chains. These factors collectively reduce the net spontaneous magnetic moment, leading to a transition from very weak ferromagnetism to a moderately ferromagnetic state. This trend is further illustrated in Fig. 2.9, which shows the correlation between increasing BT content and the evolution of magnetic properties. Notably, the introduction of BT into the BFO matrix not only moderates the magnetization but also reduces magnetic hysteresis loss, a desirable characteristic for applications such as magnetic recording media where moderate magnetization and low energy loss are critical.

Among the studied compositions, the 0.8BFO–0.2BT sample (as highlighted in Fig. 2.9) demonstrated a particularly significant enhancement in magnetization, underscoring the effectiveness of this specific ratio in optimizing magnetic behavior. Furthermore, the coercivity measured for this sample exceeds the values previously reported at low temperatures (10 K) for BFO–BT systems, indicating a notable enhancement in room-temperature magnetic performance. These findings reinforce the potential of the 0.8BFO–0.2BT composition as a promising candidate for practical ME applications due to its balanced magnetic response and improved coercivity, both critical for the design of multifunctional devices.

In-depth low-temperature investigations on $La_{0.7}Sr_{0.3}MnO_3$–$PbZr_{0.52}Ti_{0.48}O_3$ (LSMO–PZT) nanocomposites were conducted by Dwivedi et al. [86], focusing on the temperature dependence of direct current (DC) magnetization. The magnetization behavior under zero-field-cooled (ZFC) and field-cooled (FC) conditions, recorded at a magnetic field strength of 100 Oe, is presented in Fig. 2.10a. In the ZFC protocol, the sample was cooled to the minimum temperature of 175 K in the absence of a magnetic field, followed by magnetization measurements during

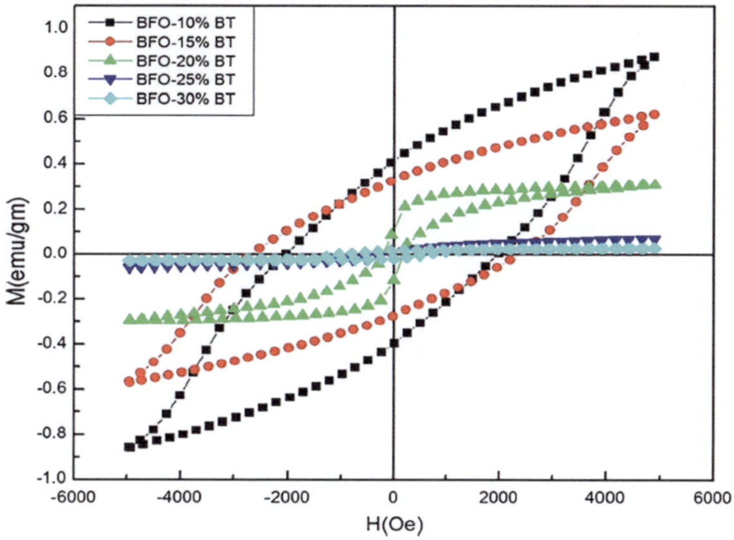

Fig. 2.9 Room temperature magnetization versus magnetic field (M–H) hysteresis loops of (1–x)BFO–xBT solid solutions, recorded to investigate the magnetic behavior as a function of BT content. Adapted with permission from Ref. [85]

warming. In contrast, the FC measurement was carried out by cooling the sample in the presence of a 100 Oe magnetic field. The resulting ZFC and FC magnetization curves exhibit a bifurcation, with the point of intersection varying across different samples (e.g., LP1, LP2, LP5). This intersection is a hallmark of superparamagnetic behavior, indicating a transition from a multi-domain FM state to a single-domain configuration.

Superparamagnetism in these nanocomposites arises when individual magnetic nanoparticles, such as LSMO, become small enough to form single-domain structures. In this state, thermal fluctuations are sufficient to spontaneously randomize the direction of magnetization, resulting in zero net magnetization in the absence of an applied field. However, under a small external field (like 100 Oe), the system exhibits a measurable magnetization similar to that of a small permanent magnet.

The shift of the bifurcation point (blocking temperature) toward higher temperatures with increasing PZT concentration suggests a suppression of interparticle interactions and enhanced superparamagnetic character. As the superparamagnetic volume fraction decreases, the thermal energy required to overcome anisotropy barriers increases, thus elevating blocking temperature. This progression highlights the impact of PZT content on the magnetic domain behavior in the composite system.

All LSMO–PZT nanocomposite samples exhibit clear ferromagnetic-to-paramagnetic transitions, as seen in the M–T plots. The transition is well-defined by a sharp change in magnetization at the Curie temperature, confirmed through the minima in the derivative plots (dM/dT vs. T) from FC measurements, as shown in Fig. 2.10b. Across all compositions, Curie temperature is approximately 360 K.

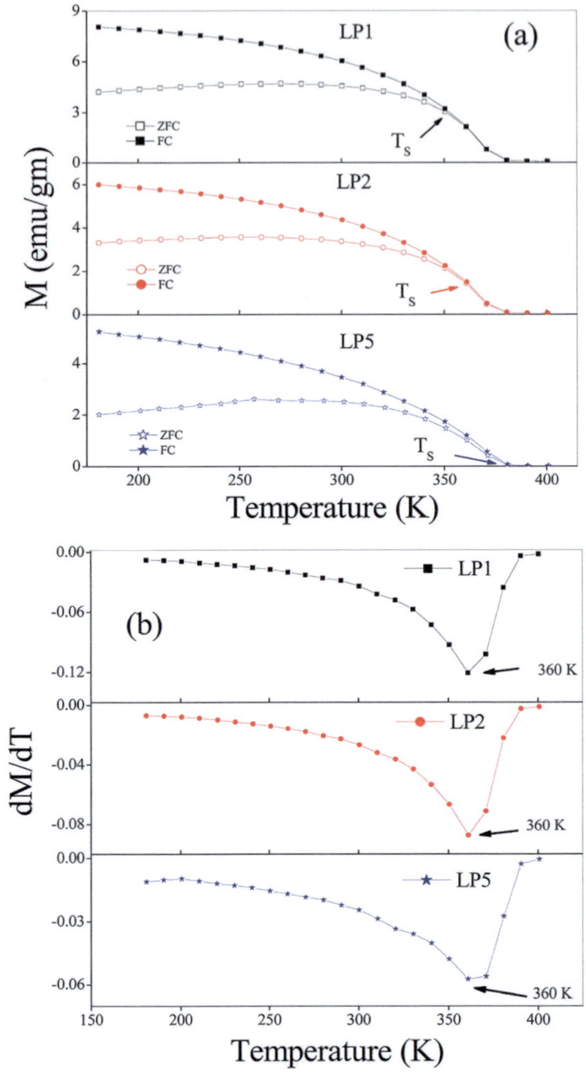

Fig. 2.10 **a** ZFC and FC magnetization versus temperature (M–T) curves for (1–*x*)LSMO–*x*PZT nanocomposites with $x =$ 0.01, 0.02, and 0.05, measured under an applied magnetic field of 100 Oe. **b** Corresponding derivative plots (dM/dT vs. T) for each composition, highlighting transition temperatures and subtle changes in magnetic behavior. Adapted with permission from Ref. [86]

Interestingly, the Curie temperature remains unchanged despite varying concentrations of PZT. This invariance suggests that PZT does not directly interact with LSMO at the atomic or electronic level. Instead, PZT occupies interfacial regions or grain boundaries within the composite, serving a structural or morphological role rather than a magnetically interactive one. As such, the intrinsic ferromagnetic properties of LSMO are preserved across the composite samples.

This separation of structural and magnetic roles highlights the importance of interfacial engineering in composite materials. The LSMO component retains its core magnetic behavior, while the inclusion of PZT allows for tunability of structural and

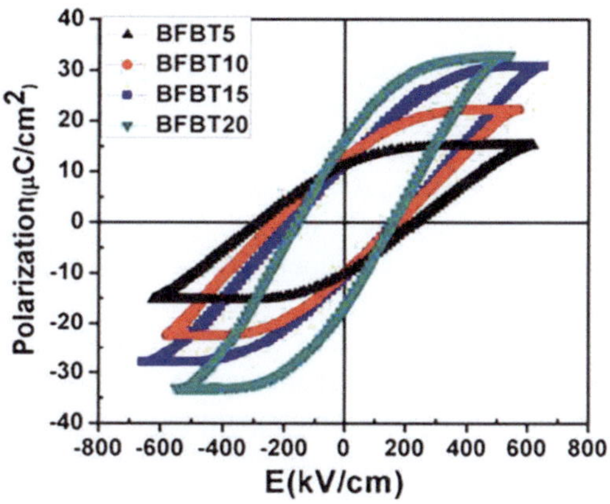

Fig. 2.11 Ferroelectric P versus E loops of BFO/BT bilayer thin films at different BFO thicknesses values under room temperature condition. Adapted with permission from Ref. [7]

possibly ferroelectric properties, without compromising magnetic performance [86]. Additionally, there are number of magnetic investigations are discussed via VSM in MF- and ME-based materials in the available latest literature [1, 85–91].

As far as the ferroelectric properties are concern, Gupta et al., explain the room temperature polarization versus electric (P vs. E) field measurements at 50 Hz on thin films of $BiFeO_3/BaTiO_3$ (BFO/BT) bilayer interface. The ferroelectric hysteresis loops are shown in Fig. 2.11 [7]. Well saturated loops are observed with a large remnant polarization $\sim$30 μC/cm^2 with coercive field $\sim$155 kV/cm for BF-20%BT. This value is maximum with respect to other reported compositions and competitive to other available values [7].

2.3.3 Dielectric, ME Coupling and Impedance Spectroscopy Characterizations

Linear fitting of the high-temperature magnetization data for various MF and ME materials reveals critical information about their magnetic transition temperature (T_o), offering valuable insight into the behavior of these materials at elevated temperatures. In particular, BFO-based compositions have shown a dielectric anomaly near the magnetic transition, suggesting a coupling between magnetic and electric orders near the paramagnetic–antiferromagnetic transition. This coupling phenomenon is well-documented across numerous BFO-based systems and underscores the presence of intrinsic ME behavior in bulk materials. For example, in $Bi_2Fe_4O_9$, both magnetic ordering and dielectric anomalies occur at a relatively low Néel temperature ($T_n \approx 240$ K). Beyond this temperature, any correlation between AFM ordering and dielectric properties disappears, with the dielectric response becoming linearly dependent on temperature. The dielectric loss factor (tan δ) also decouples from AFM characteristics above T_n.

However, in the context of BFO–BT solid solutions studied by Kumar et al. [85], $Bi_2Fe_4O_9$ plays a negligible role due to its low T_n and the significantly higher sintering (940 °C) and measurement temperatures (50–500 °C). Therefore, its influence on the magnetic and dielectric behavior in BFO–BT composites is minimal. In these BFO–BT systems, the temperature dependence of the inverse DC magnetization, recorded at field strength of 1000 Oe, is illustrated in Fig. 2.12a [91]. The data, collected across a temperature range of 50–400 °C, demonstrates that the magnetic transition temperature (T_0) for BFO–10%BT is approximately 349 °C, as determined by extrapolating the high-temperature linear region. Notably, T_0 shows a decreasing trend with increasing BT content, indicating compositional tuning of magnetic behavior. Accompanying this magnetic transition is a prominent anomaly in the dielectric constant near T_0 for BFO–10%BT, implying a strong interdependence between dielectric and magnetic properties. This correlation suggests that varying BT content influences not only magnetic ordering but also the dielectric response, indicative of robust ME coupling near the AFM–paramagnetic transition.

Further supporting this interpretation are the broad, frequency-dependent dielectric peaks observed in Fig. 2.12b. These peaks shift significantly with frequency, with the dielectric maximum temperature ($T_{o(m)}$) moving from 225 °C at 1 kHz to 375 °C at 1 MHz, a hallmark behavior of relaxor FE materials. Simultaneously, the magnitude of the dielectric constant decreases at higher frequencies. Further characterizing the relaxor nature, where response of the material is highly sensitive to the electric field frequency.

Dielectric loss behavior, as shown in Fig. 2.12c, provides additional insights into the thermal and frequency-dependent electrical performance of the material. Across the 1–1000 kHz range, dielectric loss peaks shift toward higher temperatures with increasing frequency, reinforcing the presence of relaxor-like behavior. Importantly, at 1000 kHz, the dielectric loss remains below 0.8 in the BFO–10%BT solid solution. This relatively low loss reflects both high-frequency stability and reduced energy dissipation, making the material suitable for practical applications under AC fields [91]. Moreover, solid solutions with higher BT content ($x = 0.15$ and 0.30) demonstrate a marked reduction in dielectric loss, attributed to suppressed electrical conductivity. Inclusion of BT effectively mitigates charge transport pathways and reduces dielectric losses, thus enhancing the overall electrical efficiency and functional performance of the composite [91].

In a significant parallel investigation, Singh et al. explored the MF and ME coupling behavior in BFO–BT solid solutions, emphasizing their potential for next-generation spintronic devices [92]. The study highlights that FE and ME coupling can serve as critical order parameters for multifunctional material systems. A strategic approach involving the cross-coupling of FE and magnetic components is demonstrated through the incorporation of BT into the BFO matrix, resulting in notable changes in magnetic behavior. While pure BFO typically exhibits AFM order due to its cycloidal spin structure, the introduction of BT leads to a disruption of this long-wavelength spin modulation, inducing a transition toward weak FM. This transition reflects a suppression of the cycloidal spin structure, which is widely considered

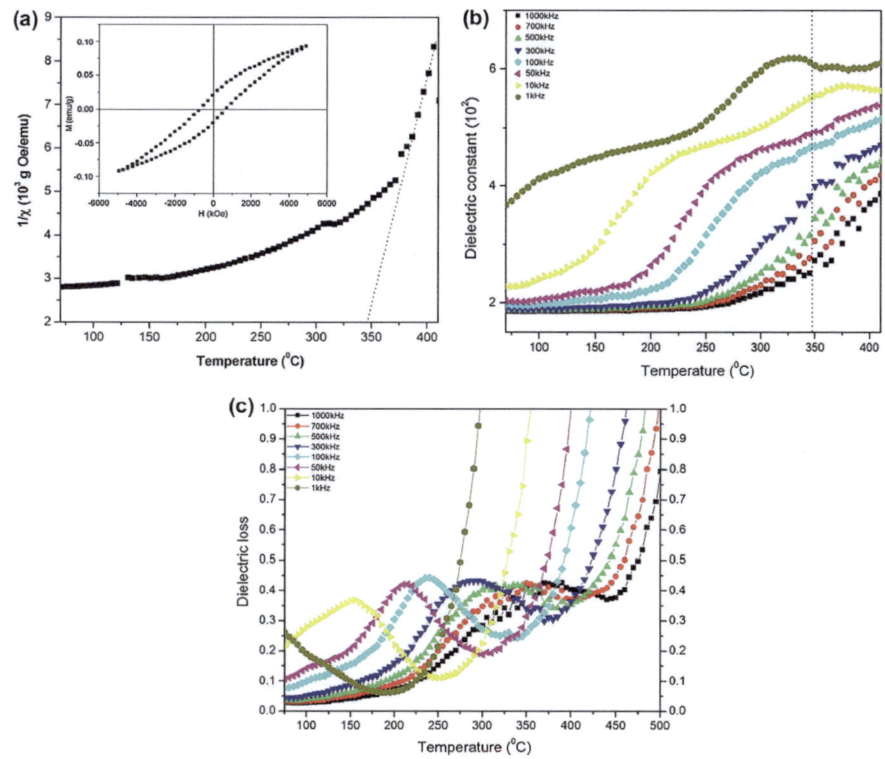

Fig. 2.12 Temperature dependence of key multifunctional properties for the $(1–x)BiFeO_2$–$xBaTiO_2$ ($x = 0.10$) solid solution. **a** Inverse magnetic susceptibility (χ^{-1}) measured at 1 kOe over the temperature range 50–400 °C, showing linear fitting in the high-temperature region to determine the magnetic transition temperature ($T_o \approx 349$ °C). Inset: Room temperature M–H hysteresis loop indicating weak ferromagnetic behavior. **b** Dielectric constant ($\varepsilon^{'}$) as a function of temperature measured at selected frequencies: 1, 10, 50, 100, 300, 500, 700, and 1000 kHz. A frequency-dependent shift in the dielectric maximum ($T_{o(m)}$) is observed, characteristic of relaxor ferroelectric behavior. **c** Dielectric loss (tan δ) as a function of temperature at the same frequencies, highlighting the frequency-dependent shift of loss peaks and low loss values (<0.8 at 1000 kHz), confirming good dielectric performance and relaxor behavior. Adapted with permission from Ref. [91]

a major barrier to achieving ME coupling in pure BFO. Consequently, the BFO–BT system displays enhanced ME interactions, attributable to this structural and magnetic reconfiguration.

The presence of ME coupling in the 0.9BFO–0.1BT composition is qualitatively demonstrated in Fig. 2.13a, which shows the temperature dependence of the inverse magnetic susceptibility ($1/\chi$) and heat flow data. The magnetic transition temperature (T_C) is observed at approximately 650 K, as indicated by the deviation in the susceptibility plot and supported by calorimetric data. This transition temperature

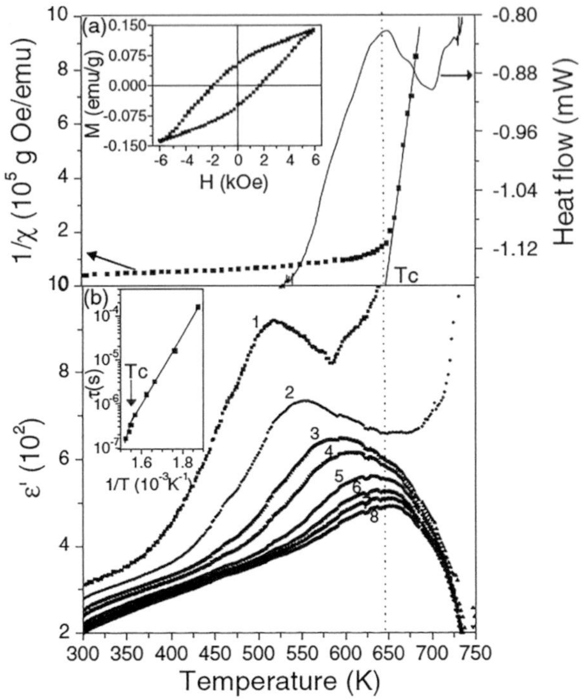

Fig. 2.13 Confirmation of the qualitative ME coupling in 09BFO-0.1BT solid solutions via. temperature dependent, **a** magnetization (dotted line) (1/v vs. T) and heat flow (continuous line), **b** dielectric measurements at some fixed frequencies and the insets represent the M-H hysteresis loop and Arrhenius plot for the polarization relaxation time, respectively. Adapted with permission from Ref. [92]

underscores the ability of the system to retain magnetic ordering at elevated temperatures. Corresponding dielectric measurements, shown in Fig. 2.13b, reveal broad, frequency-dependent dielectric maxima, further substantiating the relaxor ferroelectric nature of the material. The observed correlation between dielectric anomalies and magnetic transitions is a clear evidence of ME coupling, confirming the cooperative interaction between the electric and magnetic subsystems within the composite.

These results provide a convincing case for the 0.9BFO–0.1BT system as a candidate multiferroic material, where both magnetic and dielectric properties evolve synergistically. The alignment of magnetic and dielectric transitions reflects the intrinsic coupling mechanisms, supporting the consequence of such systems in the design of multifunctional materials for memory, logic, and sensor applications.

Building on this foundation, Dabas et al., have introduced manganese into the BFO–BT composition resulting in the creation of BMF-BT ($BiFeO_3$–Mn–$BaTiO_3$) variants [106]. These alterations have led to the observation of significantly enhanced ME coupling in the BMF-0.3BT MF compositions. The correlations found between the magnetic and dielectric transitions provide substantial evidence for the existence of ME coupling in the BFO–BT composition as shown in Fig. 2.14a–d as per the similar kind of explanation of the above. In response to these findings, our research team has modified the BFO–BT composition by adding manganese which has culminated in the formation of BMF-BT materials. This modification has resulted in an even greater degree of ME coupling within the BMF-0.3BT MF compositions [90].

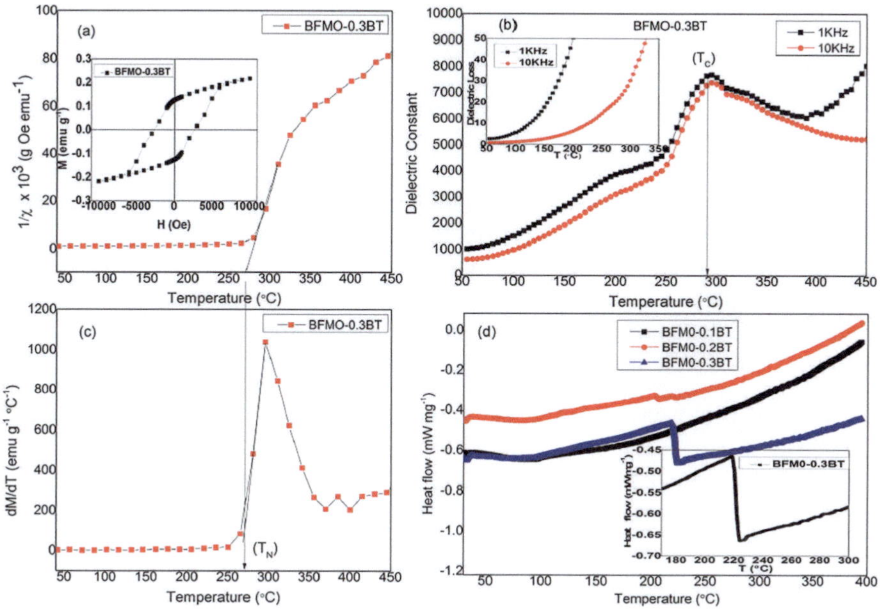

Fig. 2.14 Determination of transition temperatures and phase behavior in BFMO–*x*BT solid solutions. **a** Inverse magnetic susceptibility (1/χ) as a function of temperature for BFMO–0.3BT, with the transition temperature estimated by extrapolating the linear region of the high-temperature asymptote. Inset: M–H hysteresis loop at room temperature confirming weak ferromagnetism. **b** Temperature-dependent dielectric constant (ε') showing a broad, diffuse phase transition centered around 293 °C for BFMO–0.3BT. Inset: Dielectric loss (tan δ) versus temperature, exhibiting a corresponding relaxation peak. **c** Derivative of magnetization (dM/dT) versus temperature confirming a magnetic transition at approximately 275 °C for BFMO–0.3BT. **d** Differential scanning calorimetry (DSC) plots showing heat flow versus temperature for BFMO–0.1BT, BFMO–0.2BT, and BFMO–0.3BT compositions, indicating phase transitions. Inset: Enlarged view of the phase transition peak for BFMO–0.3BT. Adapted with permission from Ref. [90]

To complement the qualitative observations of ME coupling in BFO–BT systems, recent studies have increasingly turned toward quantitative evaluations using advanced measurement techniques. A particularly notable method is the dynamic lock-in technique, which has been employed to assess ME coupling in $BaTiO_3$–$Ni_{0.64}Zn_{0.36}Fe_2O_4$ (BT–F) MF composites [1, 93–96]. In this approach, room-temperature ME voltage measurements are conducted by applying a constant magnetic field (H_{DC}) and a superimposed alternating magnetic field (H_{AC}) across the composite sample. The resulting voltage (U_{OUT}), generated between the electrodes of the sample is directly recorded. This dynamic method not only confirms the presence of ME coupling but also enables precise evaluation of the ME response. A 1 kHz AC magnetic field is chosen to match standard dielectric characterization frequencies and align with accepted metrological practices. The ME coupling coefficient (α) is calculated using the following Eq. (2.2):

$$\alpha = \frac{1}{d} \frac{U_{OUT}}{H_{AC}} \tag{2.2}$$

where U_{OUT} is the ME-induced voltage (in mV), d is the sample thickness (in cm), and H_{AC} is the amplitude of the sinusoidal magnetic field (fixed at 4 Oe).

The variation of α as a function of H_{DC} is presented in Fig. 2.15 [1], illustrating the ME behavior of BT–F composites at different ferrite contents. The observed increase in ME voltage with rising H_{DC} is attributed to enhanced magnetostrictive strain in the ferrite phase, which is mechanically transferred to the piezoelectric matrix, thus inducing a higher voltage response. This trend demonstrates the significant role of piezomagnetic coupling in determining the ME performance. Notably, the ME coupling coefficient is strongly composition-dependent:

- For 85BT–F (15 wt% ferrite), the highest α recorded was 2.99 mV/cm Oe at 0.086 T.
- For 90BT–F, (α = 2.50 mV/cm Oe at 0.084 T.
- For 95BT–F, a much lower (α = 0.37 mV/cm Oe was observed at 0.079 T.

These values reveal a direct correlation between ferrite content, grain size, and the efficiency of ME coupling. The 85BT–F sample, possessing the largest grain size and highest interfacial contact between the magnetostrictive and piezoelectric phases, exhibited the strongest ME response. This result emphasizes the importance of microstructural engineering in optimizing ME properties.

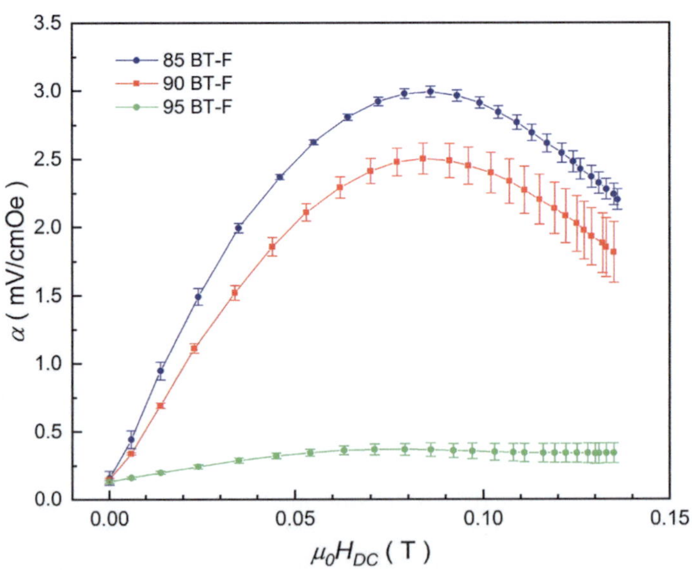

Fig. 2.15 Variation of the ME coupling coefficient (α) as a function of the constant magnetic field (H_{DC}) for BT–F MF composites. Adapted with permission from Ref. [1] under the open access license

In comparison, single-phase multiferroic materials typically show low ME coefficients, ranging from a few to several tens of mV/cm Oe. These limitations are attributed to intrinsic structural constraints. In contrast, composite and laminate structures—especially under mechanical resonance conditions—can exhibit dramatically enhanced ME responses, often exceeding hundreds of V/cm Oe. For example, a composite comprising 70%$PbZr_{0.52}Ti_{0.48}O_3$ and 30% $Ni_{0.93}Co_{0.02}Mn_{0.05}Fe_{1.95}O_{4-\delta}$ achieved an exceptional ME output of 1200 mV/cm Oe [1, 93–96]. In contrast, the 95BT–F sample, with only 5 wt% ferrite, exhibited a minimal change in ME voltage with increasing H_{DC}. This reduced response is consistent with its weaker magnetic properties noted in earlier magnetic measurements [97], highlighting the critical role of magnetostrictive phase concentration in ME coupling. The collection of ME coupling coefficient values is tabulated in Table 2.1 for single and multiphase compounds [90, 97–113].

Table 2.1 ME coupling coefficient (α) values of various compositions in single and multiphase

Sl.No	Composition	α, ME coupling coefficient ($mV\ cm^{-1}\ Oe^{-1}$)	References
1	70%$PbZr_{0.52}Ti_{0.48}O_3$–30% $Ni_{0.93}Co_{0.02}Mn_{0.05}Fe_{1.95}O_{4-\delta}$	1200	[102]
2	La_2CoMnO_6	1.45	[98]
3	$BaTiO_3$–$NiFe_2O_4$	0.5	[99]
4	$Ba_{0.9}Ca_{0.1}Ti_{0.9}Zr_{0.1}O_3$–$Co_{0.8}Ni_{0.1}Zn_{0.1}Fe_2O_4$	1.45	[100]
5	0.7$Ba_{0.95}Sr_{0.05}TiO_3$–0.3 $Ni_{0.7}Zn_{0.2}Co_{0.1}Fe_2O$	2.58	[101]
6	0.5$Na_{0.5}Bi_{0.5}TiO_3$–0.5$Co_{0.75}Zn_{0.25}Cr_{0.2}Fe_{1.8}O_4$	5.70	[102]
7	La_2NiMnO_6	≈6.8	[103]
8	$CoFe_2O_4$	47.1	[104]
9	$Zn_{0.2}Co_{0.8}Fe_2O_4$	17	[105]
10	$La_{0.7}Pb_{0.3}MnO_3$	170.1	[106]
11	$SmFeO_3$	45	[107]
12	$YFeO_3$	5.2	[108]
13	$Zn_{0.2}Mn_{0.8}Fe_2O_4$	0.16	[109]
14	$NiFe_2O_4$	90	[110]
15	BFMO-0.3BT	28.55	[90]
16	CFO/PZT thin films	238	[111]
17	$Pb(Zr_{0.52}Ti_{0.48})O_3/La_{0.67}Sr_{0.33}MnO_3$(PZT/LSMO) thin films	1153	[112]
18	Fe_3O_4	0.8	[113]

2.4 Concluding Remarks

The field of ME-MFs has seen significant advancements in recent years, driven by the development of novel synthesis methods and characterization techniques. These materials which exhibit both FE and magnetic properties hold great promise for a wide range of applications in electronics, spintronic and energy-efficient technologies. The synthesis methods discussed in this chapter including solid-state reaction, sol-gel processes, PLD, CVD, MBE, floating zone and green synthesis have enabled researchers to produce high-quality thin films, nanostructures and single crystals with precise control over composition and structure. These techniques have been crucial in enhancing the functional performance of ME-MFs by achieving epitaxial growth and minimizing impurities. Characterization techniques such as XRD, SEM, TEM, AFM, LCR meter and VSM have been instrumental in understanding the physical properties of these materials. XRD provides insights into the crystal structure and phase purity while microscopy techniques offer detailed information about surface morphology and internal structure at the nanoscale. LCR meter is essential for evaluating the dielectric properties which are crucial for the performance of ME materials in electronic applications. The exploration of ME coupling and dielectric properties has been a key focus in the field of ME-MFs. Understanding the interaction between magnetic and electric polarizations is crucial for optimizing the efficiency of these materials in various applications such as low-power memory and sensors. As the field continues to evolve, further advancements in synthesis methods and characterization techniques will undoubtedly lead to the development of more efficient and versatile ME-MFs.

References

1. Bochenek D, Niemiec P, Brzezińska D, Dercz G, Ziółkowski G, Jartych E et al (2024) Magnetoelectric properties of multiferroic composites based on $BaTiO_3$ and Nickel–Zinc ferrite material. Materials 17:1905. https://doi.org/10.3390/ma17081905
2. Gupta R, Kotnala RK (2022) A review on current status and mechanisms of room-temperature magnetoelectric coupling in multiferroics for device applications. J Mater Sci 57:12710–12737. https://doi.org/10.1007/s10853-022-07377-4
3. Kharbanda S, Dhanda N, Aidan Sun A-C, Thakur A, Thakur P (2023) Multiferroic perovskite bismuth ferrite nanostructures: a review on synthesis and applications. J Magn Magn Mater 572:170569. https://doi.org/10.1016/j.jmmm.2023.170569
4. Adnan Islam R, Priya S (2012) Progress in Dual (piezoelectric-magnetostrictive) phase magnetoelectric sintered composites. Adv Condens Matter Phys 2012:1–29. https://doi.org/10.1155/2012/320612
5. Martin LW, Crane SP, Chu Y-H, Holcomb MB, Gajek M, Huijben M et al (2008) Multiferroics and magnetoelectrics: thin films and nanostructures. J Phys: Condens Matter 20:434220. https://doi.org/10.1088/0953-8984/20/43/434220
6. Khomskii D (2009) Classifying multiferroics: mechanisms and effects. Physics (College Park Md) 2:20. https://doi.org/10.1103/Physics.2.20

7. Gupta R, Chaudhary S, Kotnala RK (2015) Interfacial charge induced magnetoelectric coupling at $BiFeO_3/BaTiO_3$ Bilayer interface. ACS Appl Mater Interfaces 7:8472–8479. https://doi.org/10.1021/am509055f
8. Ma J, Hu J, Li Z, Nan C (2011) Recent progress in multiferroic magnetoelectric composites: from bulk to thin films. Adv Mater 23:1062–1087. https://doi.org/10.1002/adma.201003636
9. Ramesh R, Spaldin NA (2007) Multiferroics: progress and prospects in thin films. Nat Mater 6:21–29. https://doi.org/10.1038/nmat1805
10. Kumar M, Shankar S, Kumar A, Anshul A, Jayasimhadri M, Thakur OP (2020) Progress in multiferroic and magnetoelectric materials: applications, opportunities and challenges. J Mater Sci: Mater Electron 31:19487–19510. https://doi.org/10.1007/s10854-020-04574-2
11. Van Aken BB, Palstra TTM, Filippetti A, Spaldin NA (2004) The origin of ferroelectricity in magnetoelectric $YMnO_3$. Nat Mater 3:164–170. https://doi.org/10.1038/nmat1080
12. Fiebig M, Lottermoser T, Fröhlich D, Goltsev AV, Pisarev RV (2002) Observation of coupled magnetic and electric domains. Nature 419:818–820. https://doi.org/10.1038/nature01077
13. Garcia-Castro AC, Ibarra-Hernandez W, Bousquet E, Romero AH (2018) Direct magnetization-polarization coupling in BaCuF4. Phys Rev Lett 121:117601. https://doi.org/10.1103/PhysRevLett.121.117601
14. Ederer C, Spaldin NA (2006) Origin of FEity in the multiferroic barium fluorides BaMF4: a first principles study. Phys Rev B Condens Matter Mater Phys 74:024102. https://doi.org/10.1103/PhysRevB.74.024102
15. Scott JF, Blinc R (2011) Multiferroic magnetoelectric fluorides: why are there so many magnetic ferroelectrics? J Phys: Condens Matter 23:113202. https://doi.org/10.1088/0953-8984/23/11/113202
16. Kimura T, Goto T, Shintani H, Ishizaka K, Arima T, Tokura Y (2003) Magnetic control of ferroelectric polarization. Nature 426:55–58. https://doi.org/10.1038/nature02018
17. Park S, Peddigari M, Hwang G-T, Yoon W-H, Kumar A, Ryu J. Face-shear 36-mode magnetoelectric composites with piezoelectric single crystal and Metglas laminate. Appl Phys Lett 115. https://doi.org/10.1063/1.5120092
18. Ghara S, Suard E, Fauth F, Tran TT, Halasyamani PS, Iyo A et al (2017) Ordered aeschynite-type polar magnets RFeWO6 (R = Dy, Eu, Tb, and Y): a new family of type-II multiferroics. Phys Rev B 95:224416. https://doi.org/10.1103/PhysRevB.95.224416
19. Liu M, Lin L, Zhang Y, Li S, Huang Q, Garlea VO et al (2017) Cycloidal magnetism driven FEity in double tungstate $LiFe(WO_4)_2$. Phys Rev B 95:195134. https://doi.org/10.1103/PhysRevB.95.195134
20. Chen R, Wang JF, Ouyang ZW, He ZZ, Wang SM, Lin L et al (2018) Magnetic field induced FEity and half magnetization plateau in polycrystalline $R_2V_2O_7$ (R = Ni, Co). Phys Rev B 98:184404. https://doi.org/10.1103/PhysRevB.98.184404
21. Ducrot É, He M, Yi G-R, Pine DJ (2017) Colloidal alloys with preassembled clusters and spheres. Nat Mater 16:652–657. https://doi.org/10.1038/nmat4869
22. Du Y, Cheng ZX, Wang XL, Liu P, Dou SX. Magnetic and ferroelectric properties of multiferroic Bi_2NiMnO_6 nanoparticles. J Appl Phys 109. https://doi.org/10.1063/1.3537943
23. Seshadri R, Hill NA (2001) Visualizing the role of Bi 6s "Lone Pairs" in the off-center distortion in ferromagnetic $BiMnO_3$. Chem Mater 13:2892–2899. https://doi.org/10.1021/cm010090m
24. Kuroiwa Y, Aoyagi S, Sawada A, Harada J, Nishibori E, Takata M et al (2001) Evidence for Pb–O Covalency in Tetragonal PbTiO3. Phys Rev Lett 87:217601. https://doi.org/10.1103/PhysRevLett.87.217601
25. Hajra S, Tripathy A, Panigrahi BK, Choudhary R (2019) Development and excitation performance of lead free electronic material: Eu and Fe doped $Bi_{0.5}Na_{0.5}TiO_3$ for filter application. Mater Res Express 6:076304. https://doi.org/10.1088/2053-1591/ab149b
26. Shimakawa Y, Azuma M, Ichikawa N (2011) Multiferroic compounds with double-perovskite structures. Materials 4:153–168. https://doi.org/10.3390/ma4010153
27. Belik AA, Azuma M, Saito T, Shimakawa Y, Takano M (2005) Crystallographic features and Tetragonal phase stability of $PbVO_3$, a new member of $PbTiO_3$ family. Chem Mater 17:269–273. https://doi.org/10.1021/cm048387i

28. Sahu RK, Subohi O (2024) Investigations on structural, dielectric and multiferroic properties of Bi_2FeNiO_6 double perovskite synthesized using sol-gel modified combustion technique. J Solgel Sci Technol 111:766–782. https://doi.org/10.1007/s10971-024-06486-7
29. Bhorde A, Waykar R, Rondiya SR, Nair S, Lonkar G, Funde A et al (2021) Structural, electronic, and optical properties of lead-free halide double perovskite Rb2AgBiI6: a combined experimental and DFT study. ES Mater Manuf. https://doi.org/10.30919/esmm5f1042
30. Hur N, Park S, Sharma PA, Ahn JS, Guha S, Cheong S-W (2004) Electric polarization reversal and memory in a multiferroic material induced by magnetic fields. Nature 429:392–395. https://doi.org/10.1038/nature02572
31. Aepuru R, Gaikwad VM, Udayabhaskar R, Panda HS, Ramalinga VM (2020) Enhanced dielectric properties and relaxation behavior in double perovskite-polymer-based flexible 0–3 nanocomposite films. J Mater Sci: Mater Electron 31:13477–13486. https://doi.org/10.1007/s10854-020-03902-w
32. Khan A, Chatterjee S, Nath TK, Taraphder A (2021) Defect-induced modulation of magnetic, electronic, and optical properties of the double-perovskite oxide La2CoMnO6. Phys Rev B 104:035152. https://doi.org/10.1103/PhysRevB.104.035152
33. Das R, Alagarsamy P, Choudhary RNP (2021) Studies of structural, electrical, and magnetic characteristics of double perovskite ceramic: La_2FeMnO_6. Phys Status Solidi (b) 258. https://doi.org/10.1002/pssb.202100299
34. Kock I, Edler T, Mayr SG (2008) Growth behavior and intrinsic properties of vapor-deposited iron palladium thin films. J Appl Phys 103. https://doi.org/10.1063/1.2875306
35. Alam M, Mandal K, Khan GG (2016) Double perovskite Y2NiMnO6 nanowires: high temperature ferromagnetic–ferroelectric multiferroic. RSC Adv 6:62545–62549. https://doi.org/10.1039/C6RA10861J
36. Farzin YA, Babaei A, Ataie A (2020) Low-temperature synthesis of Sr_2FeMoO_6 double perovskite; structure, morphology, and magnetic properties. Ceram Int 46:16867–16878. https://doi.org/10.1016/j.ceramint.2020.03.264
37. Wu H, Pei Z, Xia W, Lu Y, Leng K, Zhu X (2020) Structural, magnetic, dielectric and optical properties of double-perovskite Bi_2FeCrO_6 ceramics synthesized under high pressure. J Alloys Compd 819:153007. https://doi.org/10.1016/j.jallcom.2019.153007
38. Ravi S, Ponraj C (2013) Synthesis and characterization of new Bi_2FeNiO_6 material using a citric acid assisted gel combustion technique. Int J Mater Res 104:210–215. https://doi.org/10.3139/146.110844
39. Sai N, Kolpak AM, Rappe AM (2005) Ferroelectricity in ultrathin perovskite films. Phys Rev B 72:020101. https://doi.org/10.1103/PhysRevB.72.020101
40. Setter N, Damjanovic D, Eng L, Fox G, Gevorgian S, Hong S et al (2006) Ferroelectric thin films: review of materials, properties, and applications. J Appl Phys 100. https://doi.org/10.1063/1.2336999
41. Fong DD, Stephenson GB, Streiffer SK, Eastman JA, Auciello O, Fuoss PH et al (1979) Ferroelectricity in ultrathin perovskite films. Science 2004(304):1650–1653. https://doi.org/10.1126/science.1098252
42. Wu M, Jena P (2018) The rise of two-dimensional van der Waals ferroelectrics. WIREs Comput Mol Sci 8. https://doi.org/10.1002/wcms.1365
43. Wu M, Burton JD, Tsymbal EY, Zeng XC, Jena P (2012) Multiferroic materials based on organic transition-metal molecular nanowires. J Am Chem Soc 134:14423–14429. https://doi.org/10.1021/ja304199x
44. Lu C, Wu M, Lin L, Liu J-M (2019) Single-phase multiferroics: new materials, phenomena, and physics. Natl Sci Rev 6:653–668. https://doi.org/10.1093/nsr/nwz091
45. Tu Z, Wu M (2019) 2D diluted multiferroic semiconductors upon intercalation. Adv Electron Mater 5. https://doi.org/10.1002/aelm.201800960
46. Yang Q, Zhong T, Tu Z, Zhu L, Wu M, Zeng XC (2019) Design of single-molecule multiferroics for efficient ultrahigh-density nonvolatile memories. Adv Sci 6. https://doi.org/10.1002/advs.201801572

47. Li L, Wu M (2017) Binary compound bilayer and multilayer with vertical polarizations: two-dimensional ferroelectrics, multiferroics, and nanogenerators. ACS Nano 11:6382–6388. https://doi.org/10.1021/acsnano.7b02756
48. Qiu Z-C, Zhou J-P, Zhu G, Chen X-M, Yang R-L, Song Y-H et al (2012) Hydrothermal synthesis of perovskite bismuth ferrite crystallites with the help of NH_4Cl. J Nanosci Nanotechnol 12:6552–6557. https://doi.org/10.1166/jnn.2012.5435
49. Chauhan M, Arya E, Vibha, Kumari A, Sanghi S, Agarwal A (2023) Structural, magnetic, dielectric, magneto-electric and ferroelectric properties of $BaTiO_3$–$NiFe_2O_4$ multiferroics. J Alloys Compd 965:171360. https://doi.org/10.1016/j.jallcom.2023.171360
50. Zhang H, Kajiyoshi K (2010) Hydrothermal synthesis and size-dependent properties of multiferroic bismuth ferrite crystallites. J Am Ceram Soc 93:3842–3849. https://doi.org/10.1111/j.1551-2916.2010.03953.x
51. Sharma Y, Agarwal R, Collins L, Zheng Q, Ievlev AV, Hermann RP et al (2020) Self-assembled room temperature multiferroic $BiFeO_3$–$LiFe_5O_8$ nanocomposites. Adv Funct Mater 30. https://doi.org/10.1002/adfm.201906849
52. Wang Y, Xu G, Yang L, Ren Z, Wei X, Weng W et al (2009) Hydrothermal synthesis of single-crystal bismuth ferrite nanoflakes assisted by potassium nitrate. Ceram Int 35:1285–1287. https://doi.org/10.1016/j.ceramint.2008.04.016
53. Bhargava GK, Bhardwaj S, Singh M, Batoo KM (eds) (2021) Ferrites and multiferroics. Springer, Singapore. https://doi.org/10.1007/978-981-16-7454-9
54. Rhaman MM, Matin MA, Hossain MN, Mozahid FA, Hakim MA, Islam MF (2019) Bandgap engineering of cobalt-doped bismuth ferrite nanoparticles for photovoltaic applications. Bull Mater Sci 42:190. https://doi.org/10.1007/s12034-019-1871-8
55. Siddique M, Khan NM, Saeed M (2019) Photocatalytic activity of bismuth ferrite nanoparticles synthesized via sol-gel route. Z Phys Chem 233:595–607. https://doi.org/10.1515/zpch-2018-1225
56. Wang N, Xue C, Wan F, Zhao Y, Xu G, Liu Z et al (2020) Spontaneously conversion from film to high crystalline quality stripe during molecular beam epitaxy for high Sn content GeSn. Sci Rep 10:6161. https://doi.org/10.1038/s41598-020-63152-y
57. Lüth H (2015) Solid surfaces, interfaces and thin films. Springer International Publishing, Cham. https://doi.org/10.1007/978-3-319-10756-1
58. Ptak AJ (2015) Principles of molecular beam epitaxy. Handbook of crystal growth. Elsevier, pp 161–92. https://doi.org/10.1016/B978-0-444-63304-0.00004-4
59. Mohammad SN (2020) Synthesis of nanomaterials, vol 307. Springer International Publishing, Cham. https://doi.org/10.1007/978-3-030-57585-4
60. Bossini D, Juraschek DM, Geilhufe RM, Nagaosa N, Balatsky AV, Milanović M et al (2023) Magnetoelectrics and multiferroics: theory, synthesis, characterisation, preliminary results and perspectives for all-optical manipulations. J Phys D Appl Phys 56:273001. https://doi.org/10.1088/1361-6463/acc8e1
61. Carter CB, Norton MG (2013) Ceramic materials. Springer, New York, NY. https://doi.org/10.1007/978-1-4614-3523-5
62. Zhang TW, Mao ZW, Gu ZB, Nie YF, Pan XQ (2017) An efficient and reliable growth method for epitaxial complex oxide films by molecular beam epitaxy. Appl Phys Lett 111. https://doi.org/10.1063/1.4990663
63. Sun L, Yuan G, Gao L, Yang J, Chhowalla M, Gharahcheshmeh MH et al (2021) Chemical vapour deposition. Nat Rev Methods Primers 1:5. https://doi.org/10.1038/s43586-020-00005-y
64. Saeed M, Alshammari Y, Majeed SA, Al-Nasrallah E (2020) Chemical vapour deposition of graphene—Synthesis, characterisation, and applications: a review. Molecules 25:3856. https://doi.org/10.3390/molecules25173856
65. Yang X, Zhang G, Prakash J, Chen Z, Gauthier M, Sun S (2019) Chemical vapour deposition of graphene: layer control, the transfer process, characterisation, and related applications. Int Rev Phys Chem 38:149–199. https://doi.org/10.1080/0144235X.2019.1634319

66. Sun Z, Yan Z, Yao J, Beitler E, Zhu Y, Tour JM (2010) Growth of graphene from solid carbon sources. Nature 468:549–552. https://doi.org/10.1038/nature09579
67. Yao Y, Li Z, Lin Z, Moon K-S, Agar J, Wong C (2011) Controlled growth of multilayer, few-layer, and single-layer graphene on metal substrates. J Phys Chem C 115:5232–5238. https://doi.org/10.1021/jp109002p
68. Morber JR, Ding Y, Haluska MS, Li Y, Liu JP, Wang ZL et al (2006) PLD-assisted VLS growth of aligned ferrite nanorods, nanowires, and nanobelts synthesis, and properties. J Phys Chem B 110:21672–21679. https://doi.org/10.1021/jp064484i
69. Martin LW, Chu Y-H, Ramesh R (2010) Advances in the growth and characterization of magnetic, ferroelectric, and multiferroic oxide thin films. Mater Sci Eng R Rep 68:89–133. https://doi.org/10.1016/j.mser.2010.03.001
70. (2018) Metal oxide-based thin film structures. Elsevier. https://doi.org/10.1016/C2015-0-06152-4
71. Huijben M, Martin LW, Chu Y-H, Holcomb MB, Yu P, Rijnders G et al (2008) Critical thickness and orbital ordering in ultrathin $La_{0.7}Sr_{0.3}MnO_3$ films. Phys Rev B 78:094413. https://doi.org/10.1103/PhysRevB.78.094413
72. Li T, Hu Z, Zhang M, Li K, Yu D, Yan H (2012) Frequency dependence of magnetoelectric effect in epitaxial $La_{0.7}Sr_{0.3}MnO_3/BaTiO_3$ bilayer film. Appl Surf Sci 258:4558–4562. https://doi.org/10.1016/j.apsusc.2012.01.027
73. Fina I, Dix N, Rebled JM, Gemeiner P, Martí X, Peiró F et al (2013) The direct magnetoelectric effect in ferroelectric–ferromagnetic epitaxial heterostructures. Nanoscale 5:8037. https://doi.org/10.1039/c3nr01011b
74. Verma VK, Singh VR, Ishigami K, Shibata G, Harano T, Kadono T et al (2014) Origin of enhanced magnetoelectric coupling in $NiFe_2O_4/BaTiO_3$ multilayers studied by x-ray magnetic circular dichroism. Phys Rev B 89:115128. https://doi.org/10.1103/PhysRevB.89.115128
75. Sharma S, Tomar M, Kumar A, Puri NK, Gupta V (2014) Multiferroic properties of $BiFeO_3/BaTiO_3$ multilayered thin films. Phys B Condens Matter 448:125–127. https://doi.org/10.1016/j.physb.2014.03.089
76. Katsui H, Shiga K, Tu R, Goto T (2013) Crystal growth of $BaTi_2O_5$ by the floating zone method. J Cryst Growth 384:66–70. https://doi.org/10.1016/j.jcrysgro.2013.09.017
77. Costa FM, Ferreira NM, Rasekh S, Fernandes AJS, Torres MA, Madre MA et al (2015) Very large superconducting currents induced by growth tailoring. Cryst Growth Des 15:2094–2101. https://doi.org/10.1021/cg5015972
78. Soreto Teixeira S, Ferreira R, Carvalho J, Ferreira NM (2024) Unveiling the synthesis of strontium ferrites by sol-gel and laser floating zone methods for energy application. Crystals (Basel) 14:550. https://doi.org/10.3390/cryst14060550
79. Manzoor A, Afzal AM, Umair M, Ali A, Rizwan M, Yaqoob MZ (2015) Synthesis and characterization of Bismuth ferrite ($BiFeO_3$) nanoparticles by solution evaporation method. J Magn Magn Mater 393:269–272. https://doi.org/10.1016/j.jmmm.2015.05.066
80. Pattnaik SP, Behera A, Martha S, Acharya R, Parida K (2018) Synthesis, photoelectrochemical properties and solar light-induced photocatalytic activity of bismuth ferrite nanoparticles. J Nanopart Res 20:10. https://doi.org/10.1007/s11051-017-4110-5
81. Chaudhuri A, Mitra S, Mandal M, Mandal K (2010) Nanostructured bismuth ferrites synthesized by solvothermal process. J Alloys Compd 491:703–706. https://doi.org/10.1016/j.jallcom.2009.11.049
82. Chandel S, Thakur P, Thakur A (2020) Low leakage current density and improved dielectric behavior of $BiFexO_3$ nano-ceramics. J Alloys Compd 845:156287. https://doi.org/10.1016/j.jallcom.2020.156287
83. Gour A, Jain NK (2019) Advances in green synthesis of nanoparticles. Artif Cells Nanomed Biotechnol 47:844–851. https://doi.org/10.1080/21691401.2019.1577878
84. Hussain I, Singh NB, Singh A, Singh H, Singh SC (2016) Green synthesis of nanoparticles and its potential application. Biotechnol Lett 38:545–560. https://doi.org/10.1007/s10529-015-2026-7

85. Kumar M, Shankar S, Thakur OP, Ghosh AK (2015) Studies on magnetoelectric coupling and magnetic properties of (1–x)$BiFeO_3$–$x$$BaTiO_3$ solid solutions. J Mater Sci: Mater Electron 26:1427–1434. https://doi.org/10.1007/s10854-014-2557-z
86. Dwivedi GD, Kumar M, Shahi P, Barman A, Chatterjee S, Ghosh AK (2015) Low temperature magnetic and transport properties of LSMO–PZT nanocomposites. RSC Adv 5:30748–30757. https://doi.org/10.1039/C5RA04101E
87. Puli VS (Narayanabhatla), Pradhan DK, Hilton A, Katiyar RS, Ghosh K, Srinivasan G et al (2023) Magnetoelectric multiferroic properties of $BaTiO_3$–$CoFe_2O_4$–$BaTiO_3$ tri-layer thin films fabricated by pulsed laser deposition. J Magn Magn Mater 583:171061. https://doi.org/10.1016/j.jmmm.2023.171061
88. Catalan G, Béa H, Fusil S, Bibes M, Paruch P, Barthélémy A et al (2008) Fractal dimension and size scaling of domains in thin films of multiferroic $BiFeO_3$. Phys Rev Lett 100:027602. https://doi.org/10.1103/PhysRevLett.100.027602
89. Kumar MM, Srinivas A, Suryanarayana SV (2000) Structure property relations in $BiFeO_3$/$BaTiO_3$ solid solutions. J Appl Phys 87:855–862. https://doi.org/10.1063/1.371953
90. Dabas S, Kumar M, Chaudhary P, Thakur OP (2019) Enhanced magneto-electric coupling and energy storage analysis in Mn-modified lead free $BiFeO_3$–$BaTiO_3$ solid solutions. J Appl Phys 126. https://doi.org/10.1063/1.5119291
91. Kumar M, Shankar S, Kotnala RK, Parkash O (2013) Evidences of magneto-electric coupling in BFO–BT solid solutions. J Alloys Compd 577:222–227. https://doi.org/10.1016/j.jallcom.2013.04.044
92. Singh A, Pandey V, Kotnala RK, Pandey D (2008) Direct evidence for multiferroic magnetoelectric coupling in 0.9$BiFeO_3$–0.1$BaTiO_3$. Phys Rev Lett 101:247602. https://doi.org/10.1103/PhysRevLett.101.247602
93. Jartych E, Pikula T, Kowal K, Dzik J, Guzdek P, Czekaj D (2016) Magnetoelectric effect in ceramics based on bismuth ferrite. Nanoscale Res Lett 11:234. https://doi.org/10.1186/s11671-016-1436-3
94. Hrib LM, Caltun OF (2011) Effects of the chemical composition of the magnetostrictive phase on the dielectric and magnetoelectric properties of cobalt ferrite–barium titanate composites. J Alloys Compd 509:6644–6648. https://doi.org/10.1016/j.jallcom.2011.03.121
95. Priya S, Islam R, Dong S, Viehland D (2007) Recent advancements in magnetoelectric particulate and laminate composites. J Electroceram 19:149–166. https://doi.org/10.1007/s10832-007-9042-5
96. Eerenstein W, Mathur ND, Scott JF (2006) Multiferroic and magnetoelectric materials. Nature 442:759–765. https://doi.org/10.1038/nature05023
97. Ramana MV, Reddy NR, Siva Kumar K V., Murthy VRK, Murty BS (2011) Magneto-electric effect in multiferroic $Ni_{0.93}Co_{0.02}Mn_{0.05}Fe_{1.95}O_{4-\delta}$/$PbZr_{0.52}Ti_{0.48}O_3$ particulate composites: dielectric, piezoelectric properties. Modern Phys Lett B 25:345–358. https://doi.org/10.1142/S0217984911025742
98. Bhadra D, Masud MdG, De SK, Chaudhuri BK (2013) Observation of large magnetodielectric and direct magnetoelectric behavior in LCMO/PVDF 0–3 nanocomposites. Appl Phys Lett 102. https://doi.org/10.1063/1.4793307
99. Grigalaitis R, Vijatović Petrović MM, Bobić JD, Dzunuzovic A, Sobiestianskas R, Brilingas A et al (2014) Dielectric and magnetic properties of $BaTiO_3$–$NiFe_2O_4$ multiferroic composites. Ceram Int 40:6165–6170. https://doi.org/10.1016/j.ceramint.2013.11.069
100. Wang Y, Pu Y, Tian Y, Li X, Wang Z, Shi Y et al (2017) Enhanced magnetoelectric properties of the laminated $Ba_{0.9}Ca_{0.1}Ti_{0.9}Zr_{0.1}O_3$/$Co_{0.8}Ni_{0.1}Zn_{0.1}Fe_2O_4$ composites. J Alloys Compd 696:1307–1313. https://doi.org/10.1016/j.jallcom.2016.11.242
101. Sharma R, Singh V, Kotnala RK, Tandon RP (2015) Investigation on the effect of ferrite content on the multiferroic properties of $(1-x)$ $Ba_{0.95}Sr_{0.05}TiO_3$–$(x)$$Ni_{0.7}Zn_{0.2}Co_{0.1}Fe_2O_4$ ceramic composite. Mater Chem Phys 160:447–455. https://doi.org/10.1016/j.matchemphys.2015.05.019
102. Wadgane SR, Alone ST, Karim A, Vats G, Shirsath SE, Kadam RH (2019) Magnetic field induced polarization and magnetoelectric effect in $Na_{0.5}Bi_{0.5}TiO_3$–$Co_{0.75}Zn_{0.25}Cr_{0.2}Fe_{1.8}O_4$

multiferroic composite. J Magn Magn Mater 471:388–393. https://doi.org/10.1016/j.jmmm.2018.10.011
103. Tiwary S, Kuila S, Sahoo MR, Barik A, Ghosh R, Babu PD et al (2020) La_2NiMnO_6/poly(vinylidene fluoride) nanocomposites with enhanced magnetoelectric voltage. J Appl Phys 127. https://doi.org/10.1063/1.5140710
104. Mu X, Zhang H, Zhang C, Yang S, Xu J, Huang Y et al (2021) Poly(vinylidene fluoride-trifluoroethylene)/cobalt ferrite composite films with a self-biased magnetoelectric effect for flexible AC magnetic sensors. J Mater Sci 56:9728–9740. https://doi.org/10.1007/s10853-021-05937-8
105. Chakraborty S, Mandal SK, Saha B (2019) Magneto-optic and magneto-electric effects in poly(vinylidene fluoride)—$Zn_{0.2}Co_{0.8}Fe_2O_4$ nanocomposite organic flexible film. Ceram Int 45:14851–14858. https://doi.org/10.1016/j.ceramint.2019.04.216
106. Koner S, Deshmukh P, Ahlawat A, Singh R, Majumder SK, Satapathy S (2022) Effect of interface coupling between polarization and magnetization in $La_{0.7}Pb_{0.3}MnO_3$ (LPMO)/P(VDF-TrFE) flexible nanocomposite films. J Mater Sci 57:7621–641. https://doi.org/10.1007/s10853-022-07100-3
107. Ahlawat A, Satapathy S, Shirolkar MM, Li J, Khan AA, Deshmukh P et al (2018) Tunable magnetoelectric nonvolatile memory devices based on $SmFeO_3$/P(VDF-TrFE) nanocomposite films. ACS Appl Nano Mater 1:3196–3203. https://doi.org/10.1021/acsanm.8b00401
108. Sasmal A, Sen S, Chelvane JA, Arockiarajan A (2023) PVDF based flexible magnetoelectric composites for capacitive energy storage, hybrid mechanical energy harvesting and self-powered magnetic field detection. Polymer (Guildf) 281:126141. https://doi.org/10.1016/j.polymer.2023.126141
109. Martins P, Kolen'ko YV, Rivas J, Lanceros-Mendez S (2015) Tailored magnetic and magnetoelectric responses of polymer-based composites. ACS Appl Mater Interfaces 7:15017–15022. https://doi.org/10.1021/acsami.5b04102
110. Mayeen A, Kala MS, Jayalakshmy MS, Thomas S, Philip J, Rouxel D et al (2019) Flexible and self-standing nickel ferrite–PVDF-TrFE cast films: promising candidates for high-end magnetoelectric applications. Dalton Trans 48:16961–16973. https://doi.org/10.1039/C9DT02856K
111. Koner S, Deshmukh P, Ahlawat A, Sagdeo A, Singh R, Karnal AK, et al (2021) Effect of magnetic field on ferroelectric output voltage: a study on $La_{0.7}Sr_{0.3}MnO_3$ (LSMO)/P(VDF-TrFE) flexible multiferroic nanocomposite films. J Mater Sci: Mater Electron 32:21780–21797. https://doi.org/10.1007/s10854-021-06699-4
112. Sim CH, Pan AZZ, Wang J (2008) Thickness and coupling effects in bilayered multiferroic $CoFe_2O_4/Pb(Zr_{0.52}Ti_{0.48})O_3$ thin films. J Appl Phys 103. https://doi.org/10.1063/1.2940014
113. Zhou P, Liang K, Liu Y, Qi Y, Ma Z, Zhang T (2018) Thickness dependence of magnetoelectric properties for $Pb(Zr_{0.52}Ti_{0.48})O_3/La_{0.67}Sr_{0.33}MnO_3$ heterostructures. Ceram Int 44:12905–12908. https://doi.org/10.1016/j.ceramint.2018.04.101

Chapter 3
Sustainable Multiferroic and Magnetoelectric Materials for Energy Harvesting Applications

3.1 Introduction

The global community is seeking sustainable solutions to the critical challenges of the energy crisis and environmental pollution [1, 2]. Agricultural, textile, industrial, transportation, and construction sectors are just a few examples of activities that depend on energy in modern society. Energy demand also catalyzes growth, potentially improving the quality of life in developing nations. With the rapid increase in fossil fuel consumption and the associated environmental damage, there is an urgent need for efficient and sustainable energy sources to tackle these critical challenges facing humanity [1, 3–5]. Efficient photocatalytic technologies are exceptional candidates for society's most pressing issues, particularly environmental pollution and the demand for clean energy [6–8]. Non-conventional energy sources may provide the most effective solutions to the energy crisis and pollution [9]. Among the various forms of renewable energy, the use of solar radiation for contaminant reduction is in particularly high demand [10, 11]. The rapid growth of the human population has led to a concerning increase in worldwide energy consumption. The increasing global demand for sustainable and renewable energy sources has led to the exploration of advanced materials that can efficiently convert various forms of energy into electricity. Among these, ME and MF materials have garnered significant attention due to their unique coupling properties, which allow for the simultaneous manipulation of electric and magnetic orders [12]. Many researchers have carried out research for high ME coupling effect in different MF materials, mainly composite systems and doped systems. Composite-based perovskite ferroelectric (FE) (ABO_3) type and spinel structure ferromagnetic (FM) AB_2O_4 type include $BiFeO_3$ (BFO)-$CoFe_2O_4$ (CFO) [13–15], $BaTiO_3$ (BT)/CFO [16–18], $Pb(Ni_{1/3}Nb_{2/3})O_3$–$PbTiO_3$-$PbZrO_3$/CFO [19], and CFO/$SrTiO_3$. Further, polymer-based FM composites include $Ni_{0.5}Co_{0.5}Fe_2O_4$ (NCF)/PANI [20], PVDF/$BiFe_2O_4$ [21], and Terfenol-D/PVDF–lead zirconate titanate/PVDF [22]. Also, magnetic alloy-based FE composites

M. Kumar and S. Dabas, *Multiferroic and Magnetoelectric Materials*, Materials Horizons: From Nature to Nanomaterials, https://doi.org/10.1007/978-981-95-4391-5_3

include BT–FeCoNi alloy (BTO-FCN) [23] and Tb–Dy–Fe alloy/lead–zirconate–titanate [24]. Doped materials include Ca–Ti co-doped BFO [25], La- and Ni-doped $SrFe_{12}SrFe_{12}O_{19}$ [26], Co, Ru, and La Ion-doped $KBiFe_2O_5$ [27].

ME and MF materials are advanced functional materials that exhibit a coupling between their magnetic and electric properties [28, 29]. In ME materials, an external magnetic field can induce an electric polarization, and, conversely, an electric field can influence magnetic properties of the material [12, 30]. MFs, a subset of these materials, possess more than one primary ferroic order, such as ferromagnetism, ferroelectricity, and ferroelasticity, within the same phase [31]. The coupling between these different order parameters in MFs leads to the ME effect, which is crucial for energy harvesting applications [12, 32]. The magnetic–electric–elastic couplings in MFs are illustrated in Fig. 3.1 [33]. The ability to convert mechanical, magnetic, or electric energy into usable electrical energy is one of the key properties that make ME and MF materials attractive for energy harvesting devices. These materials can harness ambient energy from sources like vibrations, magnetic fields, and light, converting it into electricity that can power small electronic devices [34]. Their high coupling efficiency between electric and magnetic fields allows for the design of compact, low-power, and highly efficient energy harvesters [35].

Furthermore, ME and MF materials can operate in a wide range of environmental conditions, making them suitable for applications in harsh or remote locations where traditional power sources are impractical. The tunability of their properties, such as resonance frequency and polarization [36, 37], through external stimuli, also enables the optimization of device performance for specific energy harvesting applications [38]. Overall, the unique combination of multifunctionality, efficiency, and adaptability in ME and MF materials positions them as essential components in developing next-generation energy harvesting technologies. Bhoi et al. 2021 have reported pore-free MF ($PbFe_{0.5}Nb_{0.5}O_3$)–($Co_{0.6}Zn_{0.4}Fe_{1.7}Mn_{0.3}O_4$) (PFN-CZFMO) composite. The measurement of ME coupling indicates that the ME behavior is due to strain-induced interactions between the FE and magnetic domains. These composites could be ideal for future low-power memory, magnetic field sensors, and various other multifunctional devices [39].

The challenge of creating materials with multiple, often incompatible, functionalities is both scientifically intriguing and practically important. Researchers aim to overcome chemical conflicts where one crystal structure supports a specific property, while another demands a different arrangement. Engineers find this multifunctional materials valuable for developing innovative technologies, for instance, highly sought-after materials like high permittivity–permeability combinations, insulating ferromagnets, and efficient thermoelectrics [40, 41]. Given the rising energy demands of information technologies, ME-MFs stand out as promising materials for reducing the energy consumption of microelectronic devices. Their applications extend from sensors and actuators to advanced memory devices [30].

Photocatalysis is a process that utilizes light energy to initiate chemical reactions, inspired by the natural process of photosynthesis, making it a promising method for clean energy production [42, 43]. Integrating photocatalysis with ME and MF materials offers a synergistic pathway for enhanced energy harvesting [29]. These

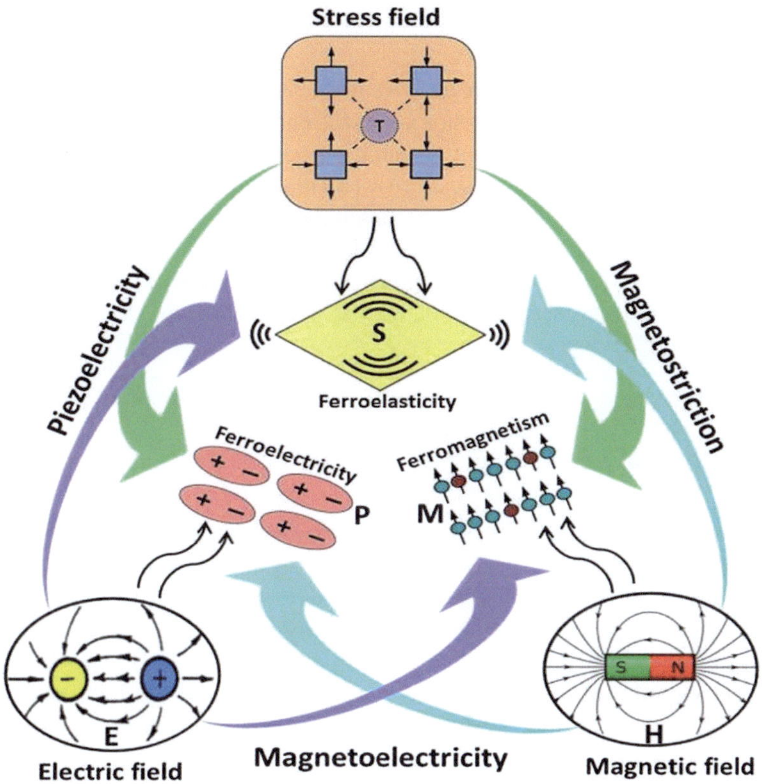

Fig. 3.1 Diagram showing the interactions between magnetic, elastic, and electric properties in MF materials. In this illustration, M represents magnetization, S denotes mechanical strain, and P indicates dielectric polarization. Adapted with permission under open access license from Ref. [33]

advanced materials can potentially overcome the limitations of conventional photocatalysts by exploiting their intrinsic coupling properties. For instance, the coupling between ferroelectricity and magnetism in MFs can induce internal electric fields that facilitate the separation and migration of photo-generated charges, thus enhancing photocatalytic efficiency [32]. Additionally, by tuning the properties of the materials through external magnetic or electric fields, it is possible to optimize the light absorption and charge transport characteristics, leading to more efficient energy conversion. In 2024, Fayaz et al. reported that the electric field-driven ionic control of functional oxides presents a flexible and efficient strategy for modifying material properties such as insulator-metal transitions, superconductivity, magnetism, and optical behavior by manipulating spin, orbit, charge, and lattice interactions. This method enables real-time tuning of these properties through external stimuli like electric fields. The potential to alter material characteristics through ionic processes is highly promising for both scientific research and practical uses, due to its energy efficiency and suitability for room-temperature operation. Initially focused on energy storage, this approach has now expanded to diverse applications in optoelectronics, nanoelectronic memory, and computing [44].

This chapter delves into the recent advancements in ME and MF materials with a focus on their applications in photocatalytic energy harvesting. It explores the underlying mechanisms and recent research on portable electronic material-based ME materials. We are focused on the electric and magnetic coupling in the material to harness power. Herein, we have also discussed the working principle and resonance conditions of ME cantilever-based energy harvester. We have also discussed the synergistic effects of ME and photocatalytic processes. Finally, we have reported the recent research advancements on integrated systems of ME and photocatalysis for energy harvesting devices.

3.2 Photocatalysis

Photocatalysis is a process that leverages light energy to drive chemical reactions [45]. In the context of energy harvesting, photocatalysis typically involves the use of semiconductor materials that can absorb photons and generate electron–hole pairs [46–48]. When light is absorbed by a photocatalyst, the energy from the photons excites electrons from the valence band to the conduction band, leaving behind positively charged holes in the valence band [7, 49]. These excited electrons and holes can then participate in redox reactions, which are fundamental to various energy conversion processes, such as splitting water to generate hydrogen, or degrading organic pollutants for environmental clean-up [8, 50].

In photocatalytic energy harvesting, the efficiency of the process depends on several factors, including the ability of the material to absorb a wide spectrum of light, the separation and migration of the photo-generated electron–hole pairs, and the prevention of recombination of these pairs before they can be utilized in chemical reactions [42, 51, 52]. Conventional photocatalysts, such as titanium dioxide (TiO_2), have been widely studied, but they often suffer from limitations such as a wide bandgap that restricts absorption to the UV region of the spectrum and rapid recombination of electron–hole pairs, which reduces the overall efficiency of the photocatalytic process [53, 54]. Various MF materials have been studied over time for photocatalysis and efforts have also been made to improve their photocatalytic properties. In 2023, Khan et al. discovered that $GdCrO_3$ nano-assemblies possess a higher intrinsic activity than IrO_2, leading to enhanced (oxygen evolution reaction) OER electrocatalytic performance. Additionally, the material exhibited a notable photocatalytic hydrogen evolution rate of up to 2 mmol/h/g_{cat}, highlighting its substantial capacity for hydrogen production [55]. The enhanced capacity of hydrogen production can improve hydrogen energy harvesting systems such as photocatalyst fuel cells [56, 57], hydrogen storage systems [58], photocatalytic water splitting systems [59], and hydrogen electrolysis systems [60]. In 2017, Arya et al. demonstrated that co-doping Sm and Co into the BFO lattice substantially improved its magnetic properties, resulting in a saturation magnetization of 8.4 emu/g, which is approximately

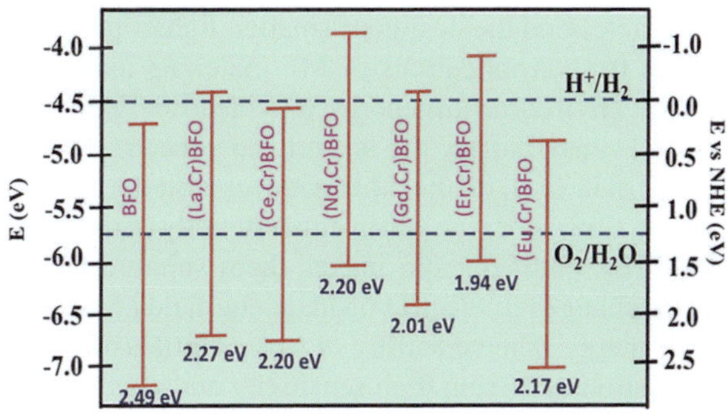

Fig. 3.2 The positions of the valence band maxima and conduction band minima for both pristine and co-doped BFO, calculated and aligned concerning the water redox potentials. Adapted with permission from Ref. [62]

20 times greater than that of pure BFO (0.40 emu/g). Additionally, the photocatalytic activity of Sm1Co1 was significantly enhanced, achieving around 92% degradation of methylene blue dye under visible light irradiation [61]. The photocatalytic properties of Sm1Co1 were examined in relation to its semiconducting FE characteristics. This analysis could provide valuable insights for creating more effective photocatalysts for solar energy-driven applications.

In 2023, Ahmad et al. explored the potential for photocatalytic activity by analyzing band edge alignment relative to the water redox potential levels. They studied structural, electronic, optical, and photocatalytic attributes of pure and (RE, Cr) co-doped BFO where (RE = Nd, Gd, La, Ce, Er, Eu) in view of density functional theory calculations. They found that Er- and Cr-doped BFO exhibited the lowest bandgap of 1.94 eV, compared to 2.51 eV for undoped BFO, without forming any mid-gap states. Among the doped samples, Gd:BFO and Nd:BFO displayed the highest polarization and magnetic moments, respectively. The low photocatalytic efficiency of BFO was attributed to the conduction band minimum (CBM) being below the water reduction potential. This limitation was shown to be improved through co-doping. The calculated positions of the valence band maxima and conduction band minima for pristine and co-doped BFO, aligned with respect to the water redox potentials, are presented in Fig. 3.2 [62].

3.3 Electric Powering for Portable Electronic Devices

The increasing demand for portable electronic devices has spurred interest in advanced materials that can provide efficient and compact power sources [63]. Among these materials, MF and ME materials stand out due to their unique ability to couple electrical, magnetic, and sometimes optical properties [64]. MFs, which exhibit simultaneous FE and FM properties, enable the coupling of electric and magnetic fields, which can be harnessed for power generation [28, 65]. For example, in these materials, an applied electric field can induce changes in magnetization, and vice versa, through ME coupling. This effect allows for high energy conversion efficiencies and the design of smaller, lighter power systems ideal for portable electronics [65]. Applications of MF materials include energy harvesting devices that convert environmental energy into electrical power and power supplies for remote sensors where battery life is a critical concern [66–68].

ME materials, on the other hand, exhibit the ability to convert magnetic fields into electric signals or vice versa, either directly or through indirect coupling mechanisms [69]. This characteristic makes them suitable for energy conversion applications where changes in electric or magnetic fields need to be translated into usable electrical energy. The versatility of ME materials allows them to be tailored for specific applications, offering high sensitivity and efficient power generation. They are used in portable energy sources such as miniature generators and energy harvesters, as well as in active devices like actuators and sensors.

Deng et al. introduced an ME antenna device, where the receiver system comprises in-plane series sensor consisting crystal fibers of Mn-PMNT and Metglas. This sensor demonstrates exceptional magnetic field detection capabilities, achieving a sensitivity of 3.1 $pT/Hz^{1/2}$ at 1 Hz. For the transmitting end, the authors fabricated "asymmetric" ME antenna device that comprised of PZT-5H and Metglas. This transmitter exhibits two resonance peaks obtained in very low-frequency (VLF) region, with radiation strengths of 73 nT and 6.3 nT at 14 kHz and 5.63 kHz, respectively. Additionally, a nonlinear magnetostrictive model was obtained, and corresponding ME effects were analyzed using finite element analysis. The model revealed that the observed near-field radiation performance of the fabricated device is majorly facilitated by frequency and "magnetic bias." Here, the ME composite-based device performs optimally at resonant frequency characterized by the ideal piezomagnetic coefficient [70]. In 2021, Wu et al. introduced a self-charging MF module designed to sustainably power personal mobile electronics. This innovation leverages MF composites to harness biomechanical energy through a mechano-magneto-electric (MME) conversion process. The energy harvesting device comprises (i) adjustable permanent magnet that transforms mechanical energy into magnetic energy and (ii) pair of piezoelectric/magnetostrictive ME laminate structure which further converts magnetic energy into usable electrical energy. The device demonstrates impressive MME energy conversion efficiency, generating an open-circuit voltage of approximately 17 V and a short-circuit current of around 7.2 μA during mechanical excitation, akin to the energy generated by human running. The module has been successfully tested as a power source device for use in sustainable personal electronics and possibly the Internet of Things [71]. Zhao et al. have developed ultra-flexible and malleable MF heterostructures composed of iron (Fe) and $BaTiO_3$. These structures exhibit an FM resonance field shift of 690 Oe in the out-of-plane direction when flexibly tuned to a radius of 5 mm. This research is expected to promulgate the development of flexible spintronic and functional devices that offer high speed, portability, and low energy consumption [72].

Recently, Arshad et al. introduced an eco-friendly, lead-free series of $BaTi_{1-x}Gd_xO_3$ ($0 \leq x \leq 0.06$) nanostructures synthesized using the sol–gel autocombustion method. The study revealed that Gd doping increases the leakage current density due to defects formed in the $BaTiO_3$ matrix, which also affects the FE properties of the materials. FE analysis showed a decrease in polarization parameters for the Gd-doped samples. The energy storage efficiency was found to be 81% and 83% for the compositions $x = 0.02$ and $x = 0.04$, respectively. The linear increase was observed for magnetic moment as well as the dielectric constant for doped samples thereby making the system highly suitable for potential multifunctional device applications [73]. Sasmal et al. [74] synthesized $Bi_{0.95}Ba_{0.05}Fe_{0.95}Zr_{0.05}O_3$ (BBFZO) nanoparticles via the sol–gel method, which were then integrated into Poly(vinylidene fluoride) (PVDF) thus facilitating the formation of nanocomposite. Measurement of the mechanical energy harvesting of the aforementioned composite films was conducted by measuring the electrical output generated through repeated human finger tapping. The 15BBFZO composite showed a significant increase in average open-circuit peak-to-peak voltage (V_{OC}), rising from approximately 3 V for pure PVDF to around 20 V. The impact of "corona poling" on the electrical performance of these composite films was thoroughly examined. The poled 15BBFZO exhibited an enhanced value of energy storage density of 139.54 μJ/cm^3 along with storage efficiency of 93.3%, as compared with a value of 139.54 μJ/cm^3 and 91.84% for the unpoled sample at 1 kHz with an applied field of ~14 kV/cm. Additionally, for poled 15BBFZO sample, the average V_{OC} increased to around 30V. A bridge rectifier circuit was utilized for AC voltage rectification from poled 15BBFZO sample and in process a 10 μF commercial capacitor was charged to approximately 5.5V DC. This voltage was successfully utilized to power LEDs thereby demonstrating the potential of the manufactured nanogenerator for small-sized and self-powered flexible nano-electronic device [74]. The value of energy storage density and efficiency of the synthesized composite films as a function of frequency along with the output of peak-to-peak open-circuit AC voltage generated by continuous human finger tapping onto the samples is illustrated in Fig. 3.3.

Recent advancements in research are focused on improving the performance and practicality of these materials. Developments include enhanced synthesis methods for

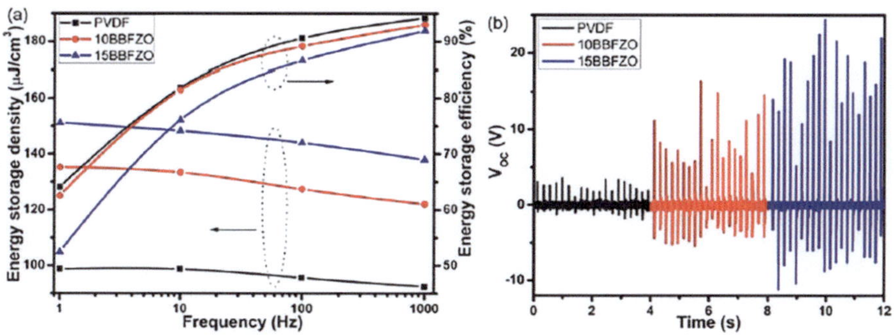

Fig. 3.3 **a** Energy storage density and efficiency of the composite films as a function of frequency. **b** Output peak-to-peak open-circuit AC voltage generated by repeated human finger tapping on the samples. Adapted with permission from Ref. [74]

producing high-quality MF and ME materials, integration techniques for embedding these materials into compact devices, and strategies for optimizing energy conversion efficiency and operational stability. The ongoing research and technological innovations continue to expand the potential applications of MF and ME materials, driving the evolution of portable electronic devices and paving the way for more effective and efficient power solutions.

3.4 ME Cantilever-Based Energy Harvester—Working Principle and Resonance Condition

ME cantilever-based energy harvester works by converting ambient magnetic energy into electrical energy through the interaction between magnetostrictive and piezoelectric materials within a cantilever structure [75–77]. The magnetostrictive material deforms under an external magnetic field, inducing mechanical stress that is transferred to the piezoelectric layer, which then generates an electric charge [78, 79]. The cantilever design amplifies this mechanical strain, enhancing the electrical output. The efficiency of the harvester is greatly increased when operating at its resonance frequency, where the frequency of the external magnetic field matches the natural frequency of the cantilever [80, 81]. At resonance, the cantilever vibrates with maximum amplitude, resulting in higher energy conversion efficiency [82]. By designing the cantilever to resonate with the expected ambient magnetic field frequencies, the ME harvester can effectively capture and convert magnetic energy, making it ideal for powering low-energy devices [83].

Guo et al. conducted a study on flexible FeCoSiB/ZnO thin film generators designed for energy harvesting device operated at low frequencies, using three different substrates for comparison. Both rigid silicon (Si) and flexible substrates, including polyethylene terephthalate (PET) and polyimide (PI), were employed in order to assess the efficiency of energy conversion. Among the tested configurations, the PET-based flexible ME generator demonstrated the highest ME coupling performance was measured in presence of an applied alternating magnetic field of 40 Oe at frequency of 20 Hz. This produced an average coupling voltage of approximately 0.643 mV and a power output of around 41.3 nW. The corresponding ME coupling coefficient for this micrometer-scale harvester reached 321.5 mV/(cm Oe). Furthermore, flexible ME modules featuring double-cantilevered structured ME generators were developed and tested. The average value of output voltage and power was found to be approximately 0.067 mV and 0.447 nW, respectively, corresponding to the series connection of the two PET-generator devices. Notably, the produced value of energy harvested for bulk MF and ME materials is considerably higher than that of the reported case of thin film laminate structured ME generators. Nevertheless, the thin film ME generator demonstrates the potential of using flexible FeCoSiB/ZnO generators for capturing ambient magnetic energy and powering sustainable electronic devices [84]. The schematic framework of the harvester and COMSOL-based magnetic field simulation corresponding to three locations of movable magnet is shown in Fig. 3.4.

In 2023, Gerhardt et al. introduced an innovative approach for creating an ME field sensor designed to be highly sensitive and reliable for biomedical applications, without the need for shielding or cryogenic cooling. The fabricated sensor comprised of piezoelectric layer with resonant cantilever along with a permanent magnet pair of which one is attached to the cantilever and another fixed to a sample holder. This approach is derived from the fundamental composite-based sensor, whereby the magnets interact in similar way as two particles in polymer matrix [85]. Empirical measurements, taken without any bias, demonstrated a detection limit of 46 $pT/\sqrt{Hz}$ and sensitivity of 2170 V/T at the sensor's resonance frequency of 223.5 Hz under ambient conditions. The sensor is fabricated using "low-resolution silicon technology" thereby ensuring high compatibility and potential integration into MEMS devices. The design of this sensor is highly adaptable and can be easily modified to suit specific application requirements. For instance, the operating resonance frequency can be adjusted not only during the cantilever's production but also by configuring the permanent magnets. Additionally, this concept can be extended to energy harvester devices as well. Beyond potential mechanical excitation, such sensor can convert a magnetic stray field of 20 μT into 1.31 $\mu W/cm^3\ Oe^2$ [86].

3.5 Synergistic Effects of ME and Photocatalytic Processes

In recent years, the integration of ME and photocatalytic processes has gained significant attention due to the potential for enhanced energy efficiency and multifunctionality in materials. The synergy between these processes arises from the unique ability of ME materials to couple magnetic and electric fields, which can be leveraged to influence and control photocatalytic reactions. In ME-based photocatalytic systems, the application of an external magnetic field can induce changes in the electric polarization within the material, leading to the modulation of charge carrier dynamics that are crucial for photocatalysis [87]. This effect can enhance the separation of photogenerated electron–hole pairs, reducing recombination rates and thereby improving the overall photocatalytic efficiency.

In 2023, Wang et al. employed the method of unit co-substitution, starting with wurtzite-type ZnO, and replaced [Zn^{2+}–Zn^{2+}] units with [A^+–M^{3+}] units, where A^+ could be Li^+, Na^+, K^+, Rb^+, Cs^+, Cu^+, Ag^+, Au^+, Pd^+, Hg^+, or Tl^+, and M^{3+} could be B^{3+}, Al^{3+}, Ga^{3+}, or In^{3+}. This approach led to the creation of a component database for the β-AMO_2 series oxides. By screening for FE photocatalytic functional materials, four promising candidates—β-$CuGaO_2$, β-$AgGaO_2$, β-$CuInO_2$, and β-$AuInO_2$—were identified. These materials were found to be potential photocatalysts for high-performance intrinsic FE semiconductors, offering phase stability, significant polarization, a narrow energy band gap, and lower effective mass, as determined by first-principles calculations [88]. Additionally, the coupling between

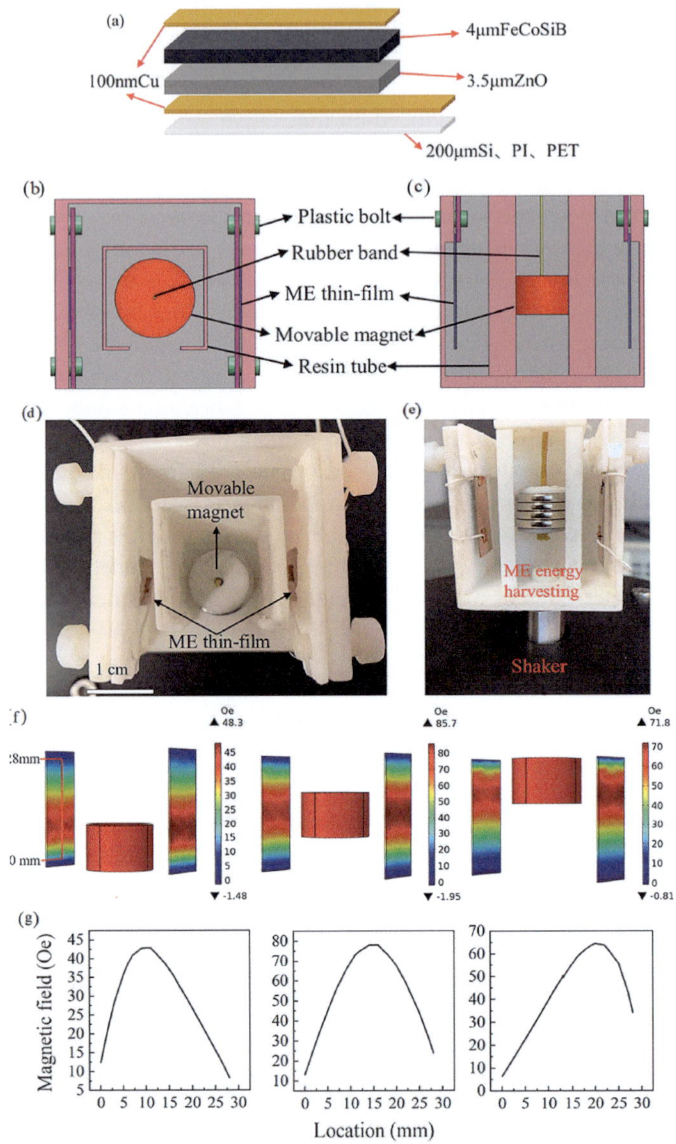

Fig. 3.4 ME material-based energy harvester device: **a** schematic of the device comprising if Si substrate over which Fe–Co–Sib and ZnO layers are sandwiched between two 100 nm-thick Cu layers. **b**, **c** Top and orthogonal view of energy harvester device. **d**, **e** Top and orthogonal view of double-cantilevered device. **f** Magnetic field simulation model developed with COMSOL where we have considered three locations of the movable magnet. **g** Variation of magnetic field as a function of location along the laminate whereby the movable magnet is considered with respect to three specific locations. Adapted with permission under open access license from Ref. [84]

magnetic and electric fields in ME materials can be tuned to optimize the light absorption properties, further boosting the photocatalytic activity under various illumination conditions. Sasmal et al. in 2023 reported $YFeO_3$-poly(vinylidene fluoride) (YFO-PVDF) composite whereby they varied the concentration of YFO in PVDF in order to corroborate the multifunctional properties including dielectric, piezoelectric, capacitive energy storage, mechanical energy harvesting, and ME-based applications. Among the composites, the film containing 5 wt% YFO (5 YF) exhibited the highest polar phase percentage of approximately 85%. In contrast, the study of energy storage performance revealed that the film with 3 wt% YFO (3 YF) had the highest energy storage density. It was observed that the 5 YF film reported highest energy harvesting performance with both the piezoelectric and piezo-tribo hybrid devices which exhibited open-circuit voltages of about 10 and 74 V and power densities of around 2 and 47 $\mu W/cm^2$, respectively. The 5 YF film exhibited significantly higher polar phase and space charge polarization. Additionally, the composite films demonstrated strong ME properties, with the poled 5 YF film showing an ME coupling coefficient (α_{33}) of approximately 5.2 mV cm^{-1} Oe^{-1} at a 1 kHz frequency and 0.8 kOe applied DC field. The application of an external magnetic field was also shown to enhance the mechanical energy harvesting performance of both the piezoelectric and piezo-tribo hybrid devices, indicating their potential for future use in self-powered magnetic field sensing applications [89].

The combination of ME effects with photocatalysis also opens up new possibilities for designing self-powered systems [90]. For instance, in a hybrid energy harvesting device, the mechanical energy from ambient vibrations or magnetic fields can be converted into electrical energy through the ME effect, which can then power the photocatalytic reactions without the need for an external power source [89, 91, 92]. This integration not only enhances the sustainability of the process but also broadens the scope of applications for ME-based materials, including in environmental remediation, water splitting [59], and hydrogen production [50, 58, 93].

Overall, the synergistic interaction between magnetoelectricity and photocatalysis represents a promising approach to developing advanced materials and devices that combine high efficiency, multifunctionality, and energy autonomy. As research in this area progresses, it is expected that new material systems and device architectures will emerge, offering even greater enhancements in performance and broadening the applicability of ME-based photocatalytic technologies.

3.6 Role of Domain Wall

A domain wall (DW) in ME is a boundary between two regions (or domains) within a material where the direction of magnetic or electric polarization differs. In ME materials, these domains can exhibit both magnetic and electric properties, which makes DWs particularly interesting for various applications, including data storage, sensors, and energy harvesting devices [94]. Significant research has been dedicated to exploring FE domain walls (DWs) in ME-MFs due to the diverse and intriguing physical phenomena they exhibit, including electrical conductivity [95–97], photovoltaic response [98, 99], magnetoresistivity [99], and spin transport [100]. DWs play a crucial role in influencing ME-MFs FE and magnetic properties. It is commonly believed that FE polarization decreases at domain walls on account of incomplete shielding of the depolarization field or the effects of gradient energy [101]. Understanding the mechanisms that enhance FE polarization and magnetic behavior at FE domain walls could advance fundamental research and multifarious potential applications in ME-MFs.

However, through collective analysis of transmission electron microscopy along with the first-principles calculations, Tao et al. in 2024 have demonstrated that the FE polarization at tail-to-tail 180° DWs in C is controlled by bound charge density. Interestingly, type-I DWs possessing low bound charge density are associated with considerably increased value, i.e., 43% of the FE polarization. While type-II DWs possessing high bound charge density have nearly null value of FE polarization. Additionally, the magnetic coupling associated with type-I and type-II FE DWs is found to be antiferromagnetic (AFM) and FM, respectively. Besides the two types of tail-to-tail domain walls observed experimentally, there are also two types of head-to-head DWs assessed through computational calculations and are also found to be energetically stable. The analysis of their FE polarization using DFT studies and FM properties are presented in Fig. 3.5. In the diagrams, the Fermi level (E_F) is represented by the dashed lines. Figure 3.5a illustrates the local density of states (LDOS) across the type-I head-to-head DW. The net magnetic moments of the two adjacent domains in the type-I DW are antiparallel, which reveals that the nature of the magnetic coupling across this DW is AFM. Figure 3.5b shows the LDOS across the type-II head-to-head DW. Here, the net magnetic moments of the two adjacent domains are parallel, indicating that the magnetic coupling across this DW is FM. These observations highlight the magnetic interactions associated with the type-I and type-II head-to-head domain walls.

In 2023, Li et al. presented research on DW ME coupling in MF hexagonal $YbFeO_3$ films. This study explored the ME coupling in h-$YbFeO_3$ thin films, where FE domain walls lead to the formation of clamped AFM DWs with diminished magnetization. Simulations based on phenomenological theory showed that the ME effect at the DWs significantly increases when the distance between FE DWs reduces to less than the characteristic width of the clamped AFM DWs during FE switching. The experimental results corroborate the unchanged value of magnetization for both reported positive and negative saturation polarization states. However, there is a reduction in the value of magnetization for coercive voltages. These outcomes suggest that the domain-wall-facilitated ME effect could effectuate electrical control of magnetization [102]. Domain-wall nanoelectronics represents an emerging blueprint for the next-generation non-volatile memory and logic device technologies, where the domain walls themselves, rather than the domains, act as the active components [103]. Particularly intriguing are charged domain walls within FE structures, characterized by sub-nanometer thicknesses and unique electronic and transport properties that hold significant potential for various nanoelectronic applications [104]. The capability to precisely generate and control charged domain walls will prove to be crucial

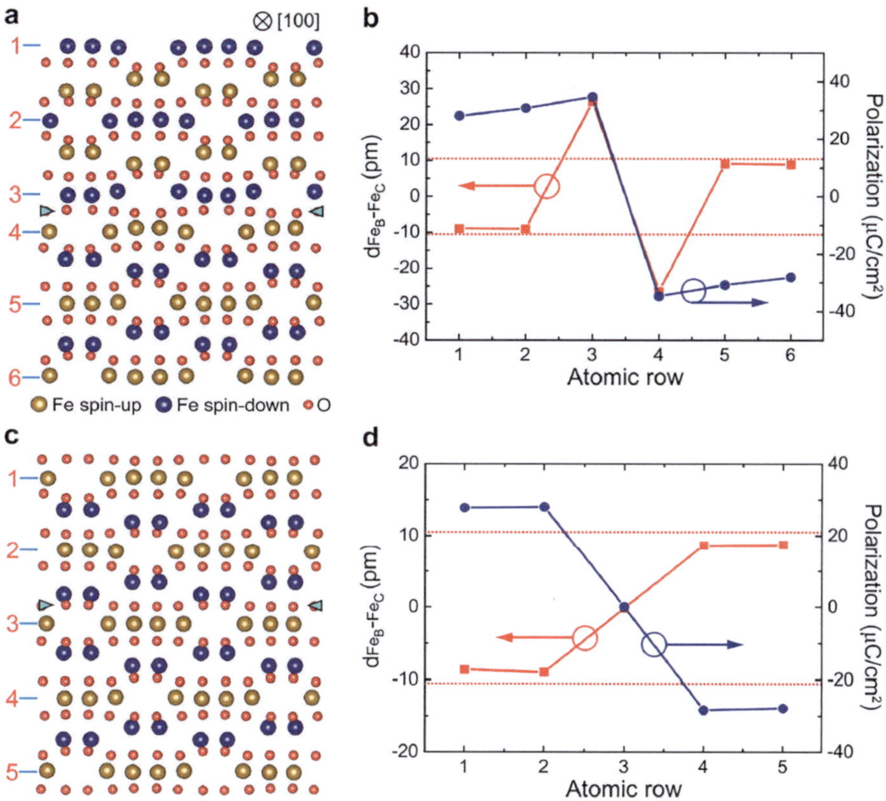

Fig. 3.5 DFT study to calculate ferroelectric polarization in two types of head-to-head ε-Fe_2O_3 DW. **a** Atomic representations of the type-I head-to-head domain wall, with the domain walls indicated by cyan arrows. **b** Calculated values of dFeB-FeC and ferroelectric polarization near the type-I head-to-head domain wall. **c** Atomic representations of the type-II head-to-head domain wall. **d** Calculated values of dFeB-FeC and ferroelectric polarization near the type-II head-to-head domain wall. The red dotted lines in **b** and **d** indicate the dFeB-FeC value of bulk ε-Fe_2O_3. Adapted with permission under open access license from Ref. [94]

in order to harness their functional properties in the next-generation portable and environmentally friendly nano-electronic devices [105].

In 2023, Liu et al. described a method for the precise creation and control of in-plane charged domain walls in thin BFO FE films, only a few nanometers thick [106]. He reported that by controlling the in-plane domain-wall position within a BFO film using voltage enables the achievement of multiple non-volatile resistance states, showcasing its essential function as a memristor with a thickness of just a few unit cells. These findings enhance the understanding of FE switching behavior and offer a novel approach for developing unit-cell-scale devices.

3.7 Integration of ME and Photocatalytic Systems for Renewable Energy Harvesting

The search for alternative clean energy sources is driven by the need to both understand their fundamental principles and develop effective technologies. Energy harvesting offers a broad opportunity to reclaim green energy from sources that are often wasted, such as solar, thermal, wind, vibration, and magnetic fields. This approach is becoming a key method for generating sustainable energy for modern electronics and wireless sensor networks. It has emerged as a vital solution to the power management challenges faced by low-power and energy-efficient devices. Utilizing sustainable FE or ME generators, energy harvesting produces electricity in a highly efficient and clean manner without releasing pollutants, making it especially suitable for "IoT"-type devices that require external power sources and regular maintenance, particularly in remote or isolated locations. By converting energy from various waste sources into electricity for powering small electronic devices, energy harvesting not only addresses battery-related issues but also helps mitigate environmental damage caused by electronic waste. This process contributes to preserving a waste-free, evergreen environment.

In his work, Saini et al., in 2024, reported a spin-charge-regulated pyro-ME nanogenerator based on the well-known FE P(VDF-TrFE) copolymer. This device effectively stores thermal-magnetic energy in a "capacitor" and generates electricity at room temperature. A schematic of the experimental setup for simultaneous thermal-magnetic energy harvesting, demonstrating the induced voltage at 2 K under thermal fluctuations with on–off H_{ac} fields and progressively increasing H_{ac} values of 0, 0.65, 1.23, and 1.65 T, is shown in Fig. 3.6a–c. A circuit for charging a capacitor using thermal-magnetic energies, along with a voltage response curve illustrating the charging process of a capacitor at 2 K under thermal fluctuations with on–off H_{ac} fields, is depicted in Fig. 3.6d, e. The FE domains are regulated to slip at the interfaces (and twins), which enhances polarization and other properties. The device achieves an excellent pyroelectric coefficient of p ~615 nC m^{-2} K^{-1}, with a notable increase in thermal sensitivity to ~1.05 V K^{-1}, an ME coefficient of α_{me} ~ 8.8 mV K^{-1} at the resonance frequency of ~180 Hz, and a magneto-sensitivity of ~473 V T^{-1}. Notably, the strategy for further improving (up to 41.2 μC m^{-2} K^{-1}) and (up to 23.6 mV cm^{-1} Oe^{-1}) has been successfully realized in electrically poled dipoles. In a model hybrid structure, spins lead to the alignment of electric dipoles parallel to the polymer chains within a cohesive charged layer. This innovative approach is highly effective for scavenging waste energy from electric vehicles, homes, and industrial sources, where thermal and magnetic energies are abundant. Such a sustainable strategy has promising potential for next-generation self-powered electronics [107].

Kim et al., in 2022, illustrated that magnetic fields can serve as an independent energy source for hydrogen production through the ME effect. Here, the MF CFO-BFO core–shell composite nanoparticles were utilized as catalysts in the hydrogen evolution reaction (HER). The HER is facilitated with the application of an AC

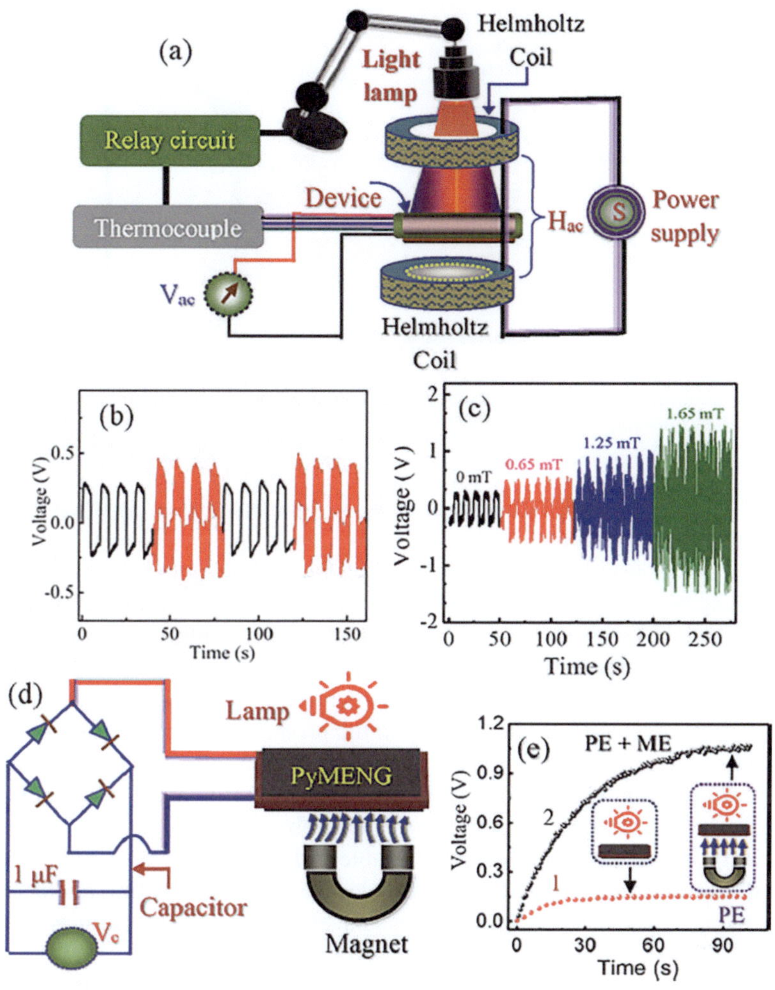

Fig. 3.6 **a** Diagram of the experimental setup for simultaneous thermal-magnetic energy harvesting, showing the induced voltage at 2 K under thermal fluctuations; **b** with on–off H_{ac} fields and **c** with progressively increasing H_{ac} values of 0, 0.65, 1.23, and 1.65 T. **d** Circuit designed for capacitor charging using thermal-magnetic energies. **e** Capacitor charging process at 2 K under thermal fluctuations with on–off H_{ac} fields. Adapted with permission from Ref. [107]

magnetic field in order to obtain an aqueous dispersion of these ME nanocatalysts. The polarization reversal driven by ME coupling generates charges on the particle surface, serving as the driving force for magnetoelectrically induced catalytic reactions. Due to the significant penetration depth of magnetic fields, even complex catalytic structures can be fully activated in opaque water environments where light is insufficient to initiate photocatalysis. These findings suggest that combining

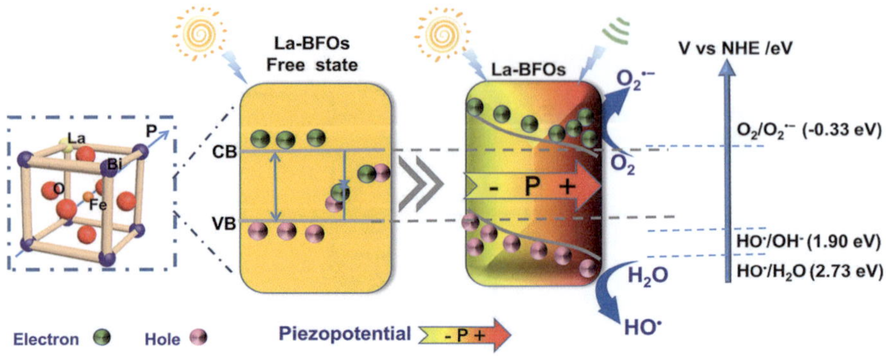

Fig. 3.7 Illustration to depict the piezoelectrically enhanced photocatalytic process in La-BFOs. Adapted with permission from Ref. [108]

magnetic fields with other energy sources could create synergistic pathways to maximize hydrogen production yield [13].

Lan et al. demonstrated significant coupling between the piezoelectric, magnetic, and photoexcitation properties of La-doped $BiFeO_3$ (La-BFO) for effective wastewater purification. They found that applying simulated solar light and ultrasonic vibrations to La-BFO particles generates a piezoelectric potential, acting as an internal electric field. This field modulates the energy band structure and accelerates the separation of photo-induced carriers, leading to the production of more reactive species that enhance the removal of carbamazepine (CBZ) from wastewater. The magnetic properties of the catalyst provide excellent catalytic stability and allow for magnetic separation during recycling experiments. Their research highlights the potential of magnetically reusable catalysts for sustainable and cost-effective water purification and demonstrates the promise of the piezoelectric effect in boosting photocatalytic activity by harnessing both solar and mechanical energies from nature as shown in Fig. 3.7 [108].

In 2023, Nien et al. investigated the photovoltaic performance of dye-sensitized solar cells (DSSCs) by modifying the photo-anode with $PbTiO_3$–$ZnFe_2O_4$/TiO_2 nanofibers to enhance current density and photovoltaic conversion efficiency (η). The flake-like $PbTiO_3$–$ZnFe_2O_4$/TiO_2 nanofibers, attached to the TiO_2 layer, achieved photovoltaic conversion efficiency of 5.87%, compared to just 4.17% for the unmodified DSSC. The improved performance is attributed to two main factors: the large amount of N719 dye adsorbed on the extensive surface area of the nanofibers and the low charge recombination rate [109].

In 2024, Khan et al. reported that $TbCrO_3$ nanoparticles exhibited a high photocatalytic hydrogen evolution rate of 10 mmol h^{-1} g^{-1}, attributed to their large specific surface area and the FE charge separation phenomena. In electrocatalytic seawater splitting, where the oxygen evolution reaction (OER) is replaced by hydrazine oxidation (HzOR), $TbCrO_3$ nanoparticles demonstrated an improved overpotential (η_{10}) of approximately 0.20 V for HzOR, compared to 0.22 V for OER versus the reversible hydrogen electrode (RHE). Additionally, the hydrogen evolution reaction in simulated water showed the generation of a significant catalytic current density. The as-synthesized MF $TbCrO_3$ nanoparticles thus exhibited multifunctional applications in both photocatalytic and electrocatalytic activities [2].

3.8 Concluding Remarks and Future Challenges

The integration of ME and MF materials into energy harvesting and powering systems for portable electronic devices marks a significant advancement in material science and engineering. These materials offer a unique combination of electrical, magnetic, and, in some cases, optical properties, making them highly suitable for applications in energy generation, storage, and conversion. The exploration of ME cantilever-based energy harvesters, synergistic effects of ME and photocatalytic processes, and the role of domain walls in enhancing material properties have opened new avenues for developing efficient, compact, and sustainable power solutions.

The research and technological innovations discussed in this chapter underscore the immense potential of these materials in revolutionizing portable electronics, renewable energy harvesting, and environmental sustainability. One of the significant challenges in the application of ME and MF materials is ensuring their long-term stability and durability, especially under varying environmental conditions. These materials are often sensitive to temperature, humidity, and mechanical stress, which can degrade their performance over time. While ME materials offer promising energy conversion capabilities, achieving high efficiency in practical applications remains a challenge. Optimizing the coupling between magnetic and electric fields, as well as improving the resonance conditions in ME harvesters, is crucial for maximizing energy output. Integrating ME materials into existing electronic devices and scaling up production for commercial applications present significant hurdles. The synthesis of high-quality, defect-free materials on a large scale, as well as the seamless integration of these materials into compact devices, requires further development. The production of some ME and MF materials involves rare or toxic elements, leading to higher costs and potential environmental concerns. Developing eco-friendly, cost-effective alternatives is necessary to make these technologies more accessible and sustainable. Future research should focus on developing novel synthesis methods that enhance the properties of ME and MF materials, such as increasing their energy conversion efficiency, stability, and flexibility. This could include exploring new compositions, doping strategies, and hybrid material systems. Continued efforts in miniaturizing energy harvesting and powering devices without compromising performance will be critical. This includes designing ME cantilever structures that operate efficiently at low frequencies and integrating ME materials into MEMS for portable applications. The synergistic integration of ME and photocatalytic processes represents a promising direction for future research. Developing systems that combine energy harvesting, storage, and environmental remediation in a single platform could lead to multifunctional devices with broad applications, including IoT and smart city technologies. The exploration of ME materials in environmental and biomedical fields (such as water purification, hydrogen production, and biosensing) offers exciting opportunities. Research into the domain-wall behavior in these materials could lead to breakthroughs in non-volatile memory devices and advanced sensors. Finally, the role of ME materials in renewable energy harvesting should be further explored. Their ability to convert waste energy from various sources into usable power can contribute to the development of sustainable energy solutions, reducing reliance on conventional power sources and mitigating environmental impact.

References

1. Hassan Q, Viktor P, Al-Musawi TJ, Ali BM, Algburi S, Alzoubi HM et al (2024) The renewable energy role in the global energy transformations. Renew Energy Focus 48:100545. https://doi.org/10.1016/j.ref.2024.100545
2. Khan H, Lofland SE, Ahmed J, Ramanujachary KV, Ahmad T (2024) Energy-efficient hydrogen generation using multiferroic $TbCrO_3$ nanoparticles via electrocatalysis coupled with hydrazine oxidation and photocatalysis. Int J Hydrog Energy 58:717–725. https://doi.org/10.1016/j.ijhydene.2024.01.257
3. Achakulwisut P, Erickson P, Guivarch C, Schaeffer R, Brutschin E, Pye S (2023) Global fossil fuel reduction pathways under different climate mitigation strategies and ambitions. Nat Commun 14:5425. https://doi.org/10.1038/s41467-023-41105-z
4. Hassan Q, Algburi S, Sameen AZ, Al-Musawi TJ, Al-Jiboory AK, Salman HM et al (2024) A comprehensive review of international renewable energy growth. Energy Built Environ. https://doi.org/10.1016/j.enbenv.2023.12.002
5. Matin M, Amin N, Zaharim A, Sopian K Bin, Matin MA, Sopian K (2010) A study towards the possibility of ultra thin Cds/CdTe high efficiency solar cells from numerical analysis. In: Renewable energy view project Malaysia renewable energy potential view project
6. Ahuja R, Srivastava R (2024) Introduction to photocatalytic materials for clean energy, renewable chemicals production, and sustainable catalysis. Nanoscale Adv 6:3695–3698. https://doi.org/10.1039/D4NA90075H
7. Náfrádi M, Veréb G, Firak DS, Alapi T (2022) Photocatalysis: introduction, mechanism, and effective parameters, pp 3–31. https://doi.org/10.1007/978-3-030-77371-7_1
8. Yang X, Wang D (2018) Photocatalysis: from fundamental principles to materials and applications. ACS Appl Energy Mater 1:6657–6693. https://doi.org/10.1021/acsaem.8b01345
9. Owusu PA, Asumadu-Sarkodie S (2016) A review of renewable energy sources, sustainability issues and climate change mitigation. Cogent Eng 3:1167990. https://doi.org/10.1080/23311916.2016.1167990
10. Singh AP, Wang RB, Tossi C, Tittonen I, Wickman B, Hellman A (2021) Hydrogen induced interface engineering in Fe_2O_3–TiO_2 heterostructures for efficient charge separation for solar-driven water oxidation in photoelectrochemical cells. RSC Adv 11:4297–4307. https://doi.org/10.1039/d0ra09655e
11. Maka AOM, O'Donovan TS (2022) Effect of thermal load on performance parameters of solar concentrating photovoltaic: high-efficiency solar cells. Energy Built Environ 3:201–209. https://doi.org/10.1016/j.enbenv.2021.01.004
12. Chen C, Zhong S, Sun G, Zhang Y, Ding Y, Ren K et al (2023) The magnetoelectric coupling effect of multiferroic fluids and their potential applications. J Mater Sci: Mater Electron 34:2041. https://doi.org/10.1007/s10854-023-11490-8
13. Kim D, Efe I, Torlakcik H, Terzopoulou A, Veciana A, Siringil E et al (2022) Magnetoelectric effect in hydrogen harvesting: magnetic field as a trigger of catalytic reactions. Adv Mater 34. https://doi.org/10.1002/adma.202110612

14. Khalid A, Saleem M, Naseem S, Ramay SM, Shaikh HM, Atiq S (2020) Magneto-electric coupling and multifunctionality in $BiFeO_3$–$CoFe_2O4$ core-shell nano-composites. Ceram Int 46:12828–12836. https://doi.org/10.1016/j.ceramint.2020.02.053
15. Sharma R, Mukherjee R, Kaur M, Adhikari SK, Nath M, Basandrai D (2024) Compositional-driven variations in magnetic, conductivity, and ferroelectric properties of multiferroic BiFeO3–CoFe2O4 composite system. J Mater Res 39:1661–1672. https://doi.org/10.1557/s43578-024-01336-2
16. Corral-Flores V, Bueno-Baqués D, Ziolo RF (2010) Synthesis and characterization of novel $CoFe_2O_4$–$BaTiO_3$ multiferroic core–shell-type nanostructures. Acta Mater 58:764–769. https://doi.org/10.1016/j.actamat.2009.09.054
17. Erdem D, Bingham NS, Heiligtag FJ, Pilet N, Warnicke P, Vaz CAF et al (2016) Nanoparticle-based magnetoelectric $BaTiO_3$–$CoFe_2O_4$ thin film heterostructures for voltage control of magnetism. ACS Nano 10:9840–9851. https://doi.org/10.1021/acsnano.6b05469
18. Etier M, Schmitz-Antoniak C, Salamon S, Trivedi H, Gao Y, Nazrabi A et al (2015) Magnetoelectric coupling on multiferroic cobalt ferrite–barium titanate ceramic composites with different connectivity schemes. Acta Mater 90:1–9. https://doi.org/10.1016/j.actamat.2015.02.032
19. Vadla SS, Kulkarni AR, Narayanan V (2016) Magnetoelectric coupling in $0.5Pb(Ni_{1/3}Nb_{2/3})O_3$–$0.35PbTiO_3$–$0.15PbZrO_3$ and $CoFe_2O_4$ based particulate composites. Scr Mater 112:140–143. https://doi.org/10.1016/j.scriptamat.2015.09.035
20. Kour S, Adhikari SK, Palihawadana M, Mukherjee R (2023) The dielectric, magnetic, and ferroelectric analysis of $xNi_{0.5}Co_{0.5}Fe_2O_4$:$(1-x)$PANI multiferroic composites. J Inorg Organomet Polym Mater 33:1704–1715. https://doi.org/10.1007/s10904-023-02602-2
21. Prabhakaran T, Hemalatha J (2012) Highly flexible poly (vinyldine fluoride)/bismuth iron oxide multiferroic polymer nanocomposites, pp 1309–1310. https://doi.org/10.1063/1.4710494
22. Lin Y, Cai N, Zhai J, Liu G, Nan C-W (2005) Giant magnetoelectric effect in multiferroic laminated composites. Phys Rev B 72:012405. https://doi.org/10.1103/PhysRevB.72.012405
23. Huang J, Li L, Hu Z, Tsai BK, Huang J, Shen J et al (2024) Ultrathin Ternary FeCoNi alloy nanoarrays in $BaTiO_3$ matrix for room-temperature multiferroic and hyperbolic metamaterial. Nano Lett 24:10081–10089. https://doi.org/10.1021/acs.nanolett.4c02036
24. Zhao G, Dong L, Quan H, Xiong C (2007) Sandwich magnetoelectric composites of polyvinylidene fluoride, Tb-Dy-Fe alloy, and lead zirconate titanate. J Wuhan Univ Technol-Mater Sci Ed 22:596–599. https://doi.org/10.1007/s11595-006-4596-7
25. Shekhar M, Pradhan LK, Kumar L, Kumar P (2023) Dielectric relaxation behavior and conduction mechanism of Ca and Ti co-doped multiferroic bismuth ferrite. J Electron Mater 52:6182–6202. https://doi.org/10.1007/s11664-023-10563-y
26. Apostolov AT, Apostolova IN, Wesselinowa JM (2024) Theoretical study of electric, dielectric, and optical properties in ion doped multiferroic $SrFe_{12}O_{19}$ nanoparticles. Materials 17:1544. https://doi.org/10.3390/ma17071544
27. Apostolov A, Apostolova I, Wesselinowa J (2023) Multiferroic properties of Co, Ru, and La Ion doped $KBiFe_2O_5$. Materials 17:1. https://doi.org/10.3390/ma17010001
28. Eerenstein W, Mathur ND, Scott JF (2006) Multiferroic and magnetoelectric materials. Nature 442:759–765. https://doi.org/10.1038/nature05023
29. Huang Y-L, Nikonov D, Addiego C, Chopdekar RV, Prasad B, Zhang L et al (2020) Manipulating magnetoelectric energy landscape in multiferroics. Nat Commun 11:2836. https://doi.org/10.1038/s41467-020-16727-2
30. Cheng Y, Peng B, Hu Z, Zhou Z, Liu M (2018) Recent development and status of magnetoelectric materials and devices. Phys Lett Sect A: Gener Atomic Solid State Phys 382:3018–3025. https://doi.org/10.1016/j.physleta.2018.07.014
31. (2023) Ferroelectricity and multiferroicity down to the atomic thickness. Nat Nanotechnol 18:829–830. https://doi.org/10.1038/s41565-023-01494-0
32. Bréhin J, Chen Y, D'Antuono M, Varotto S, Stornaiuolo D, Piamonteze C et al (2023) Coexistence and coupling of ferroelectricity and magnetism in an oxide two-dimensional electron gas. Nat Phys 19:823–829. https://doi.org/10.1038/s41567-023-01983-y

33. Palneedi H, Annapureddy V, Priya S, Ryu J (2016) Status and perspectives of multiferroic magnetoelectric composite materials and applications. Actuators 5. https://doi.org/10.3390/act5010009
34. Yu R, Feng S, Sun Q, Xu H, Jiang Q, Guo J et al (2024) Ambient energy harvesters in wearable electronics: fundamentals, methodologies, and applications. J Nanobiotechnol 22:497. https://doi.org/10.1186/s12951-024-02774-0
35. Annapureddy V, Palneedi H, Hwang G-T, Peddigari M, Jeong D-Y, Yoon W-H et al (2017) Magnetic energy harvesting with magnetoelectrics: an emerging technology for self-powered autonomous systems. Sustain Energy Fuels 1:2039–2052. https://doi.org/10.1039/C7SE00403F
36. Pradhan DK, Kumari S, Rack PD (2020) Magnetoelectric composites: applications, coupling mechanisms, and future directions. Nanomaterials 10:1–22. https://doi.org/10.3390/nano10102072
37. Ullah N, Islam MdS, Hoque A, Yong WH, Alrashdi AM, Soliman MS et al (2023) An efficient, compact, wide-angle, wide-band, and polarization-insensitive metamaterial electromagnetic energy harvester. Alex Eng J 82:377–388. https://doi.org/10.1016/j.aej.2023.10.015
38. Du K, Barkaoui H, Zhang X, Jin L, Song Q, Xiao S (2022) Optical metasurfaces towards multifunctionality and tunability. Nanophotonics 11:1761–1781. https://doi.org/10.1515/nanoph-2021-0684
39. Bhoi K, Mohanty HS, Ravikant, Abdullah MF, Pradhan DK, Babu SN et al (2021) Unravelling the nature of magneto-electric coupling in room temperature multiferroic particulate ($PbFe_{0.5}Nb_{0.5}O_3$)–($Co_{0.6}Zn_{0.4}Fe_{1.7}Mn_{0.3}O_4$) composites. Sci Rep 11. https://doi.org/10.1038/s41598-021-82399-7
40. Spaldin NA (2020) Multiferroics beyond electric-field control of magnetism. Proc R Soc A: Math Phys Eng Sci 476:20190542. https://doi.org/10.1098/rspa.2019.0542
41. Hu H, Ju Y, Yu J, Wang Z, Pei J, Thong H-C et al (2024) Highly stabilized and efficient thermoelectric copper selenide. Nat Mater 23:527–534. https://doi.org/10.1038/s41563-024-01815-1
42. Zhu S, Wang D (2017) Photocatalysis: basic principles, diverse forms of implementations and emerging scientific opportunities. Adv Energy Mater 7. https://doi.org/10.1002/aenm.201700841
43. Sakuragi Y, Bryant DA (n.d.) Genetic manipulation of quinone biosynthesis in cyanobacteria. Photosystem I. Dordrecht, Springer Netherlands, pp 205–222. https://doi.org/10.1007/978-1-4020-4256-0_15
44. Fayaz MU, Wang Q, Liang S, Bai H, Pan F, Song C (2024) Electric field induced ion migration and property tuning in functional oxides. Acc Mater Res 5:648–660. https://doi.org/10.1021/accountsmr.4c00067
45. Khan MM (2021) Principles and mechanisms of photocatalysis. In: Photocatalytic systems by design. Elsevier, pp 1–22. https://doi.org/10.1016/B978-0-12-820532-7.00008-4
46. Fujishima A, Honda K (1972) Electrochemical photolysis of water at a semiconductor electrode. Nature 238:37–38. https://doi.org/10.1038/238037a0
47. Ahmad I, Zou Y, Yan J, Liu Y, Shukrullah S, Naz MY et al (2023) Semiconductor photocatalysts: a critical review highlighting the various strategies to boost the photocatalytic performances for diverse applications. Adv Colloid Interface Sci 311:102830. https://doi.org/10.1016/j.cis.2022.102830
48. Serpone N, Emeline AV (2012) Semiconductor photocatalysis—Past, present, and future outlook. J Phys Chem Lett 3:673–677. https://doi.org/10.1021/jz300071j
49. Markushyna Y, Savateev A (2022) Light as a tool in organic photocatalysis: multi-photon excitation and chromoselective reactions. Eur J Org Chem 2022. https://doi.org/10.1002/ejoc.202200026
50. Nishiyama H, Yamada T, Nakabayashi M, Maehara Y, Yamaguchi M, Kuromiya Y et al (2021) Photocatalytic solar hydrogen production from water on a 100 m^2 scale. Nature 598:304–307. https://doi.org/10.1038/s41586-021-03907-3

51. Morshedy AS, El-Fawal EM, Zaki T, El-Zahhar AA, Alghamdi MM, El Naggar AMA (2024) A review on heterogeneous photocatalytic materials: mechanism, perspectives, and environmental and energy sustainability applications. Inorg Chem Commun 163:112307. https://doi.org/10.1016/j.inoche.2024.112307
52. Zhang J, Li Y, Zhao X, Wang L, Chen H, Wang S et al (2021) Aligning potential differences within carbon nitride based photocatalysis for efficient solar energy harvesting. Nano Energy 89:106357. https://doi.org/10.1016/j.nanoen.2021.106357
53. Rafique M, Hajra S, Irshad M, Usman M, Imran M, Assiri MA et al (2023) Hydrogen production Using TiO_2-based photocatalysts: a comprehensive review. ACS Omega 8:25640–25648. https://doi.org/10.1021/acsomega.3c00963
54. Haruna A, Chong F-K, Ho Y-C, Merican ZMA (2022) Preparation and modification methods of defective titanium dioxide-based nanoparticles for photocatalytic wastewater treatment—A comprehensive review. Environ Sci Pollut Res 29:70706–70745. https://doi.org/10.1007/s11356-022-22749-8
55. Khan H, Lofland SE, Ramanujachary KV, Alhokbany N, Ahmad T (2023) Bifunctional multiferroic $GdCrO_3$ nanoassemblies for sustainable H_2 production using electro- and photocatalysis. ACS Appl Energy Mater 6:8102–8110. https://doi.org/10.1021/acsaem.3c01210
56. Roy S, Darabdhara J, Ahmaruzzaman M (2023) Sustainable degradation of pollutants, generation of electricity and hydrogen evolution via photocatalytic fuel cells: an inclusive review. Environ Res 236:116702. https://doi.org/10.1016/j.envres.2023.116702
57. Ni J, Wen Y, Pan D, Bai J, Zhou B, Zhao S et al (2023) Light-driven simultaneous water purification and green energy production by photocatalytic fuel cell: a comprehensive review on current status, challenges, and perspectives. Chem Eng J 473:145162. https://doi.org/10.1016/j.cej.2023.145162
58. Muthukumar P, Kumar A, Afzal M, Bhogilla S, Sharma P, Parida A et al (2023) Review on large-scale hydrogen storage systems for better sustainability. Int J Hydrog Energy 48:33223–33259. https://doi.org/10.1016/j.ijhydene.2023.04.304
59. Molaei MJ (2024) Recent advances in hydrogen production through photocatalytic water splitting: a review. Fuel 365:131159. https://doi.org/10.1016/j.fuel.2024.131159
60. Nnabuife SG, Hamzat AK, Whidborne J, Kuang B, Jenkins KW (2024) Integration of renewable energy sources in tandem with electrolysis: a technology review for green hydrogen production. Int J Hydrog Energy. https://doi.org/10.1016/j.ijhydene.2024.06.342
61. Arya G, Yogiraj J, Negi NS, Shah J, Kotnala RK (2017) Observation of enhanced multiferroic, magnetoelectric and photocatalytic properties in Sm–Co codoped $BiFeO_3$ nanoparticles. J Alloys Compd 723:983–994. https://doi.org/10.1016/j.jallcom.2017.06.325
62. Ahmad T, Jindal K, Tomar M, Jha PK (2023) Effect of codoping of rare earth elements and Cr on multiferroic, optical and photocatalytic properties of $BiFeO_3$. Mater Today Commun 37. https://doi.org/10.1016/j.mtcomm.2023.107516
63. Tarascon J-M, Armand M (2001) Issues and challenges facing rechargeable lithium batteries. Nature 414:359–367. https://doi.org/10.1038/35104644
64. Zhao Y, Wang Y, Yang Y, Zhao J, Jiang X (2024) Realization of 2D multiferroic with strong magnetoelectric coupling by intercalation: a first-principles high-throughput prediction. NPJ Comput Mater 10:122. https://doi.org/10.1038/s41524-024-01301-x
65. Fang Y-W, Ding H-C, Tong W-Y, Zhu W-J, Shen X, Gong S-J et al (2015) First-principles studies of multiferroic and magnetoelectric materials. Sci Bull (Beijing) 60:156–181. https://doi.org/10.1007/s11434-014-0628-4
66. Wei X, Domingo N, Sun Y, Balke N, Dunin-Borkowski RE, Mayer J (2022) Progress on emerging ferroelectric materials for energy harvesting, storage and conversion. Adv Energy Mater 12. https://doi.org/10.1002/aenm.202201199
67. Bai Y, Jantunen H, Juuti J (2018) Energy harvesting research: the road from single source to multisource. Adv Mater 30. https://doi.org/10.1002/adma.201707271
68. Martins P, Brito-Pereira R, Ribeiro S, Lanceros-Mendez S, Ribeiro C (2024) Magnetoelectrics for biomedical applications: 130 years later, bridging materials, energy, and life. Nano Energy 126:109569. https://doi.org/10.1016/j.nanoen.2024.109569

69. Smith IT, Zhang E, Yildirim YA, Campos MA, Abdel-Mottaleb M, Yildirim B et al (2023) Nanomedicine and nanobiotechnology applications of magnetoelectric nanoparticles. WIREs Nanomed Nanobiotechnol 15. https://doi.org/10.1002/wnan.1849
70. Deng T, Lu L, Lin D, Zhu L, Bai C, Xia Z et al (2023) Acoustic actuator antenna system based on magnetoelectric laminated composites. Sens Actuators A Phys 363:114752. https://doi.org/10.1016/j.sna.2023.114752
71. Wu H, Tatarenko A, Bichurin MI, Wang Y (2021) A multiferroic module for biomechanical energy harvesting. Nano Energy 83:105777. https://doi.org/10.1016/j.nanoen.2021.105777
72. Zhao Y, Peng R, Guo Y, Liu Z, Dong Y, Zhao S et al (2021) Ultraflexible and malleable Fe/$BaTiO_3$ multiferroic heterostructures for functional devices. Adv Funct Mater 31. https://doi.org/10.1002/adfm.202009376
73. Arshad M, Khan W, Abushad M, Mohanta S, Husain S, Shukla DK et al (2024) Superior energy storage performance and excellent multiferroic properties of $BaTi_{1-x}Gd_xO_3$ ($0 \leq x \leq 0.06$) ceramics. Mater Res Bull 169. https://doi.org/10.1016/j.materresbull.2023.112504
74. Sasmal A, Sen S, Devi PS (2020) Frequency dependent energy storage and dielectric performance of Ba–Zr Co-doped $BiFeO_3$ loaded PVDF based mechanical energy harvesters: effect of corona poling. Soft Matter 16:8492–8505. https://doi.org/10.1039/d0sm01031f
75. Liu G, Ci P, Dong S (2014) Energy harvesting from ambient low-frequency magnetic field using magneto-mechano-electric composite cantilever. Appl Phys Lett 104. https://doi.org/10.1063/1.4862876
76. Garg T, Goyal LM (2023) Magnetoelectric composites-based energy harvesters. In: Piezoelectric materials—New opportunities for energy harvesting devices. IntechOpen. https://doi.org/10.5772/intechopen.110875
77. Bradai S, Naifar S, Trigona C, Baglio S, Kanoun O (2021) An electromagnetic/magnetoelectric transducer based on nonlinear RMSHI circuit for energy harvesting and sensing. Measurement 177:109307. https://doi.org/10.1016/j.measurement.2021.109307
78. Elhajjar R, Law C-T, Pegoretti A (2018) Magnetostrictive polymer composites: recent advances in materials, structures and properties. Prog Mater Sci 97:204–229. https://doi.org/10.1016/j.pmatsci.2018.02.005
79. Petrov VM, Srinivasan G, Bichurin MI, Galkina TA (2009) Theory of magnetoelectric effect for bending modes in magnetostrictive-piezoelectric bilayers. J Appl Phys 105. https://doi.org/10.1063/1.3087766
80. Wang K, Liu W, Tang Y, Pei J, Kang S, Wu Z (2023) Widening the bandwidth of vibration energy harvester by automatically tracking the resonant frequency with magnetic sliders. Sustain Energy Technol Assess 58:103368. https://doi.org/10.1016/j.seta.2023.103368
81. Kan J, Wu Y, Li S, Wang S, Zhang Z (2023) A tunable rotational energy harvester exploiting a flexible-clamping piezoelectric beam by deploying magnetic repulsive force. Sens Actuators A Phys 353:114198. https://doi.org/10.1016/j.sna.2023.114198
82. Chen L-Q, Fan Y (2023) Internal resonance vibration-based energy harvesting. Nonlinear Dyn 111:11703–11727. https://doi.org/10.1007/s11071-023-08464-0
83. Yang Z, Zhou S, Zu J, Inman D (2018) High-performance piezoelectric energy harvesters and their applications. Joule 2:642–697. https://doi.org/10.1016/j.joule.2018.03.011
84. Guo Y, Yang C, Huang B (2023) Design of flexible FeCoSiB/ZnO thin-film multiferroic module for low-frequency energy harvesting. Energies (Basel) 16. https://doi.org/10.3390/en16135049
85. Herrera-May AL, Aguilera-Cortés LA, García-Ramírez PJ, Manjarrez E (2009) Resonant magnetic field sensors based on MEMS technology. Sensors 9:7785–7813. https://doi.org/10.3390/s91007785
86. Gerhardt M, Zimoch L, Dorn C, Elzenheimer E, Bald C, Lerg T et al (2023) Self-powered elementary hybrid magnetoelectric sensor. Nano Energy 115. https://doi.org/10.1016/j.nanoen.2023.108720
87. Lv T, Li J, Arif N, Qi L, Lu J, Ye Z et al (2022) Polarization and external-field enhanced photocatalysis. Matter 5:2685–2721. https://doi.org/10.1016/j.matt.2022.06.004

88. Wang R, Xu L, Liu Q, Shi Q, Liu X (2023) Search for simple β-$A^IM^{III}O_2$-type intrinsic ferroelectric semiconductors with simultaneous robust built-in electric field and full-spectrum absorption for superior photocatalysts. J Mater Chem A Mater 11:5233–5244. https://doi.org/10.1039/D2TA09509B
89. Sasmal A, Sen S, Chelvane JA, Arockiarajan A (2023) PVDF based flexible magnetoelectric composites for capacitive energy storage, hybrid mechanical energy harvesting and self-powered magnetic field detection. Polymer (Guildf) 281:126141. https://doi.org/10.1016/j.polymer.2023.126141
90. Zhang Q, Huang C, Zhang Y, Liu S, Zhang D, Li P et al (2024) Self-powered photocatalytic system fabricated by bifunctional coordination polymers. Chem Eng J 482:148854. https://doi.org/10.1016/j.cej.2024.148854
91. Flores-Hernández DA, Palomino-Resendiz SI, Jost N, Luviano-Juárez A (2024) Numerical energy evaluation of an optimized hybrid energy harvesting system (CPV-TEG) considering the effects of solar tracking. Appl Therm Eng 247:123071. https://doi.org/10.1016/j.applthermaleng.2024.123071
92. Cheng M, Wu J, Liang X, Mao R, Huang H, Ju D et al (2023) Hybrid multi-mode magneto-mechano-electric generator with enhanced magnetic field energy harvesting performance. Sens Actuators A Phys 352:114194. https://doi.org/10.1016/j.sna.2023.114194
93. Hansora D, Yoo JW, Mehrotra R, Byun WJ, Lim D, Kim YK et al (2024) All-perovskite-based unassisted photoelectrochemical water splitting system for efficient, stable and scalable solar hydrogen production. Nat Energy 9:272–284. https://doi.org/10.1038/s41560-023-01438-x
94. Tao A, Jiang Y, Chen S, Zhang Y, Cao Y, Yao T et al (2024) Ferroelectric polarization and magnetic structure at domain walls in a multiferroic film. Nat Commun 15. https://doi.org/10.1038/s41467-024-50431-9
95. Seidel J, Martin LW, He Q, Zhan Q, Chu Y-H, Rother A et al (2009) Conduction at domain walls in oxide multiferroics. Nat Mater 8:229–234. https://doi.org/10.1038/nmat2373
96. Rojac T, Bencan A, Drazic G, Sakamoto N, Ursic H, Jancar B et al (2017) Domain-wall conduction in ferroelectric $BiFeO_3$ controlled by accumulation of charged defects. Nat Mater 16:322–327. https://doi.org/10.1038/nmat4799
97. Meier D, Seidel J, Cano A, Delaney K, Kumagai Y, Mostovoy M et al (2012) Anisotropic conductance at improper ferroelectric domain walls. Nat Mater 11:284–288. https://doi.org/10.1038/nmat3249
98. Yang SY, Seidel J, Byrnes SJ, Shafer P, Yang C-H, Rossell MD et al (2010) Above-bandgap voltages from ferroelectric photovoltaic devices. Nat Nanotechnol 5:143–147. https://doi.org/10.1038/nnano.2009.451
99. Domingo N, Farokhipoor S, Santiso J, Noheda B, Catalan G (2017) Domain wall magnetoresistance in $BiFeO_3$ thin films measured by scanning probe microscopy. J Phys: Condens Matter 29:334003. https://doi.org/10.1088/1361-648X/aa7a24
100. Daraktchiev M, Catalan G, Scott JF (n.d.) Landau theory of domain wall magnetoelectricity
101. Cao W, Cross LE (1991) Theory of tetragonal twin structures in ferroelectric perovskites with a first-order phase transition. Phys Rev B 44:5–12. https://doi.org/10.1103/PhysRevB.44.5
102. Li X, Yun Y, Thind AS, Yin Y, Li Q, Wang W et al (2023) Domain-wall magnetoelectric coupling in multiferroic hexagonal $YbFeO_3$ films. Sci Rep 13. https://doi.org/10.1038/s41598-023-28365-x
103. Catalan G, Seidel J, Ramesh R, Scott JF (2012) Domain wall nanoelectronics. Rev Mod Phys 84:119–156. https://doi.org/10.1103/RevModPhys.84.119
104. Martin LW, Rappe AM (2016) Thin-film ferroelectric materials and their applications. Nat Rev Mater 2:16087. https://doi.org/10.1038/natrevmats.2016.87
105. Sharma P, Schoenherr P, Seidel J (2019) Functional ferroic domain walls for nanoelectronics. Materials 12:2927. https://doi.org/10.3390/ma12182927
106. Liu Z, Wang H, Li M, Tao L, Paudel TR, Yu H et al (2023) In-plane charged domain walls with memristive behaviour in a ferroelectric film. Nature 613:656–661. https://doi.org/10.1038/s41586-022-05503-5

107. Saini D, Sengupta D, Mondal B, Mishra HK, Ghosh R, Vishwakarma PN et al (2024) A spin-charge-regulated self-powered nanogenerator for simultaneous pyro-magneto-electric energy harvesting. ACS Nano 18:11964–11977. https://doi.org/10.1021/acsnano.4c02406
108. Lan S, Yu C, Sun F, Chen Y, Chen D, Mai W et al (2022) Tuning piezoelectric driven photocatalysis by La-doped magnetic $BiFeO_3$-based multiferroics for water purification. Nano Energy 93:106792. https://doi.org/10.1016/j.nanoen.2021.106792
109. Nien Y-H, Wu Y-T, Chou J-C, Yang P-H, Ho C-S, Lai C-H et al (2023) Photovoltaic performance of dye-sensitized solar cell by modification of photoanode With $PbTiO_3$–$ZnFe_2O_4$/TiO_2 nanofibers. IEEE Trans Electron Devices 70:3780–3787. https://doi.org/10.1109/TED.2023.3279801

Chapter 4
Photovoltaic Performance of Sustainable Multiferroic and Magnetoelectric Materials in Green Energy Technology

4.1 Introduction

The energy crisis arises due to growing energy demand and toxicity. The various associated environmental issues forced the research community to look for the urgent green source of energy to generate electricity other than fossil fuel. With the advent of green technology, research based on the alternate and renewable energy field has been observing a frequent and considerable attention with substantial growth in recent times. The solar radiation in their wide spectrum from sun emerges out as an inexhaustible source of energy among other alternative power sources and extensively employed around the globe. The solar cell operated under the PV effect mainly depends on the materials employed, mechanisms, and the conversion efficiency well recognized to harvest the accessed solar radiation and is a potential contender to traditional source of energy [1, 2]. The first-generation semiconductor, mainly silicon-based single *p–n* junction conventional solar cell devices, is widely employed commercially to generate electricity. The upper theoretical limit of the power conversion efficiency of the first-generation solar cell devices can be 31% which can be further enhance to maximum 41% depending on the various factors including the concentration of incident solar radiation [3, 4]. The crystalline silicon-based solar cell module is known for delivering the relatively high efficiency and its fully developed fabrication process hugely dominates the solar cell commercial production. However, the leading issue in harvesting the solar energy from the PV with respect to the present fossil fuel technology is the high cost involved in manufacturing the silicon wafer which is employed in the solar cell.

However, the second-generation solar cell primarily centered on designing thin film of various semiconducting materials other than crystalline silicon deposited on the cost-effective substrate such as glass for developing single-junction or multi-junction device [5]. Though the manufacturing cost of this thin film-based solar cell devices is relatively low but experiences the same performance limitation as the previous one. The key factor for developing the cost-effective second generation

M. Kumar and S. Dabas, *Multiferroic and Magnetoelectric Materials*, Materials Horizons: From Nature to Nanomaterials, https://doi.org/10.1007/978-981-95-4391-5_4

of solar cell devices mainly includes employing the cheap and suitable substrate, less material consumption, and large-scale manufacturing [5]. The multiple layers of different and unique bandgap semiconductor in the multi-junction solar cells provides the absorption of incident light from the respective range of the solar radiation to generate the electricity. The multi-junction solar cell leads to further improvement in the PV performances and overcome the SQ limit. The inherent complexities involved in developing the multi-junction devices lead to further increase the undesirable cost of production. The second-generation solar cell constituted multi-junction has shown very impressive and highest reported efficiency of ~45.7%. However, the efficiency of infinite-junction solar cell calculated computationally is reported to be ~87% [5].

The merit of high power conversion efficiency (PCE) and the low cost of the first- and second-generation solar cell, respectively, are directed to put together for the evolution of third-generation PV solar cell. In essence, the third-generation solar cell primarily includes organic/polymeric/plastic solar cells, dye-sensitized solar cells, quantum dot solar cells, nanocrystal-based solar cells, and perovskite solar cells (PSCs). The cost-effective technology, employing non-toxic, environment-friendly, advanced, and novel materials for efficient light absorption, and finally the suitable device architecture are to be combined together for the further progress and development of the third-generation solar cells in order to compete with the present-day market of silicon-based solar cells [4, 5]. In view of this, few FE and MF materials are successfully employed as the light-absorbing materials and give significant photo-response. These FE and MF material-based PV phenomena strengthen the belief of getting the superior PCE under the third-generation solar cell. The FE materials primarily possess significant permanent electric dipoles which exist even with no external electric field applied. The ferroelectricity is found to be very promising and appealing characteristics which enables regulation of internal electric field and injected barriers and results in the employment of FE materials to generate electricity. The illumination of the sandwiched FE materials between the respective electrodes produces the photo-excited charge carriers which are duly separated because of inherent polarization-based present internal electric field. The FE materials produced the electricity due to bulk PV effect which is entirely based on atomistic asymmetry of the current generation. The FE domains and electric polarization present in the materials are mainly responsible for separation of charge carriers generated by light irradiation and thus further result in the enhancement of the PV activity [6–11]. However, the magnetic domain arrangement and spin–orbit interaction as part of magnetic property of the materials enable the generated photocurrent to flow. In view of this, to establish the effect of magnetic and FE properties on the PV activity, the ME materials are found to be suitable materials [12, 13].

In recent decades, MF materials have attracted much awaited attention among the researchers and scientific community worldwide on account of possessing the intriguing and still clandestine physical properties with the prospect of developing and advancing the new technological applications including the PV. A ferroic material with its inherently possessed properties enables the applied conjugate field to regulate present internal alignment/order. This internal alignment present in the ferromagnetic, FE, and ferroelastic materials made to switch spontaneously with the applied external magnetic field, electric field, and external stress, respectively. However, there are ferrotoroidic materials in which possible alignment of toroidal moments switched by applying the crossed electric and magnetic fields [14]. The MF materials with single or composite phase more often reported to exhibit more than one ferroic order parameters such as ferromagnetism, ferroelectricity, and ferroelasticity under certain conditions [15, 16]. The first known possible MF was boracites [4].

Figure 4.1a shows the inter-wind relationship existing among various ferroics leading to MF characteristics in the materials [16]. However, the MF materials demonstrate both ferroelectricity and magnetism simultaneously at room temperature which are the most promising materials. They provide an extra degree of freedom for developing the novel multifunction devices such as sensors, filters, inductors, antennas, energy harvester, storage, and PV devices [17–27]. These types of MF materials with cross-coupling effect between magnetic and electric ordering states are now well known as MF-ME materials. However, the microscopic origin of the magnetic ordering in the magnetic materials is completely different with respect to the circumstances which resulted in the phenomena of ferroelectricity in a material. The ME effect facilitated by the presence of magnetization and polarization enabled the conversion between stored energies in magnetic and electric fields. The MF-ME materials exhibit the distinctive property of modifying the electric polarization when external magnetic field is applied, however, when subjected to external electric field resulted in modified net magnetization. Figure 4.1b well represents the ME coupling and their respective hysteresis loop [17]. Based on application point of view, $TbMnO_3$ (TMO) with magnetic control polarization and $BiFeO_3$ (BFO) with large polarization are widely investigated and promising MF-ME materials have been widely explored in recent times [28, 29].

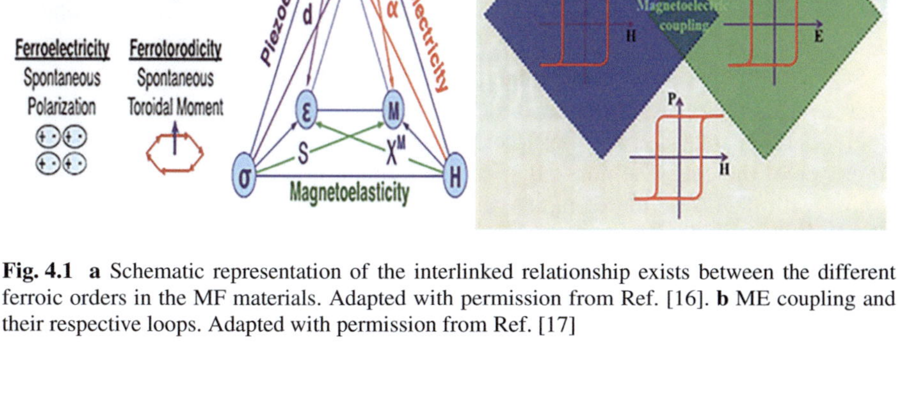

Fig. 4.1 **a** Schematic representation of the interlinked relationship exists between the different ferroic orders in the MF materials. Adapted with permission from Ref. [16]. **b** ME coupling and their respective loops. Adapted with permission from Ref. [17]

Among all single-phase multiferroic (MF) materials, $BiFeO_3$ is a distinctive magnetoelectric (ME) compound, possessing a high ferroelectric transition temperature (Curie temperature, $T_c \sim 1103$ K and a Néel transition temperature of ~643 K) [28, 29]. It also receives the great attention because of its wide band gap ~2.16 eV as it lies in the range of visible spectrum of the solar radiation and make the BFO potential candidate as a photoactive material for PV applications. The distorted rhombohedral perovskite structure of BFO with R3c symmetry space group makes it to have larger polarization at room temperature. A significant amount of photocurrent has been reported to be generated due to distorted, i.e., non-centrosymmetric structure which could further be utilized for the production of electrical energy [28, 29]. So far, the biggest challenge in employing the BFO in the PV application is to further reduce the band gap while retaining the FE characteristics.

In the present chapter, it is attempted to discuss various factors involved in deciding the upper theoretical limit of efficiency of a single-junction solar cell, known as SQ limit and briefly discussing the possible ways to overcome it. The quantum efficiency which is the ability of converting the light by PV devices is also discussed here. It is attempted to briefly discuss lead-free and environment-friendly green ME-MF materials with the mechanism involved in the ME effect. The various PV performance parameters of the recently studied green ME-MF material are also discussed.

4.2 Shockley–Queisser Limit

When the sun light is incident on a PV cell on account of PV effect, this cell converts the sun light into the electricity. The sun light on the earth surface mainly constitutes its deep-ultraviolet, visible and infrared spectrum with visible spectrum at peak range. However, the solar conversion efficiency of the PV cell is found to be limited by various factors. The William Shockley and Hans J. Queisser were the first one to calculate the efficiency limit in their research work "Detailed Balance Limit of Efficiency of *p–n* Junction Solar Cells" in 1961 [30]. The maximum solar conversion efficiency of the single *p–n* junction solar cell calculated theoretically is known as SQ limit or detailed balance limit. The amount of the converted electrical energy per incident photon becomes the main bulwark and investigated thoroughly in calculating the SQ limit [30, 31].

SQ takes into the consideration the radiative recombination mechanism involving electron–hole pair as it is mandated ideally for the detailed balance principle in the single-junction solar cell in order to calculate the maximum theoretical limit of the efficiency of the cell. The non-radiative recombination was also taken into consideration with fixed fraction of radiative recombination in calculating the efficiency. However, the sun, source of solar energy and solar cell device were assumed to be blackbodies with temperature 6000 K and 300 K, respectively. The efficiency for the purpose of evaluating its detailed balance limit is ordinarily the ratio of output electrical power to the incident solar power. However, this efficiency depends on various variables which are indispensible and even include which result in reduction of efficiency below the detailed balance limit [30].

SQ takes into account the above-mentioned factors to calculate the ultimate efficiency, current–voltage relationship, nominal efficiency, and well-known detailed balance limit of the efficiency and compares it with semi-empirical limit of the efficiency. The upper limit of the solar PCE of single *p*–*n* junction solar cell with band gap of 1.4 eV found to be ~33.7% under the solar spectrum of AM 1.5 [30]. The upper limit of the efficiency and open-circuit voltage V_{oc} and short-circuit current J_{sc} of an ideal single-junction solar cell as evaluated as function of energy band gap by the SQ is well illustrated in Fig. 4.2a–d [32].

The upper limit of the efficiency has resulted in exploring and investigating other materials and configurations as the photoactive material including the organics, polymers, forefront perovskite, and even an elaborate systematic layering of the materials such as tandem solar cells [33, 34]. There are two main strategies suggested to be the promising pathways to surpass the SQ upper limit of efficiency. First one evolved as carrier multiplication mechanism under this incident photon with energy at least two times greater than the band gap mainly absorbed by the photo-carriers. The interaction between incident photon and the photo-carriers results in activating the inverse

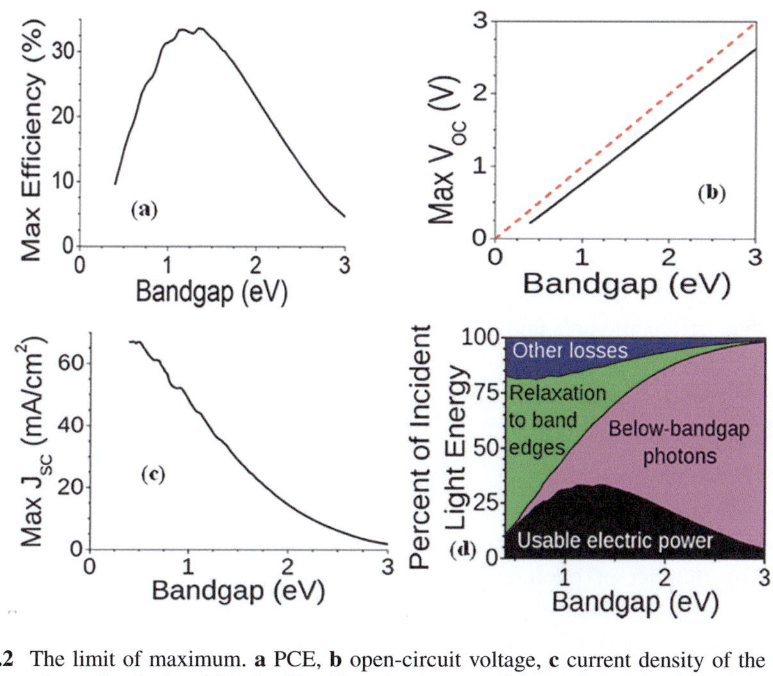

Fig. 4.2 The limit of maximum. **a** PCE, **b** open-circuit voltage, **c** current density of the single-junction solar cell calculated by the Shockley–Queisser model as a function of band gap of the semiconductor, **d** schematic illustration of percentage of incident light energy utilized for the electrical power conversion along loss with respect to band gap of the semiconductor. Adapted with permission from Ref. [31, 32]

Auger emission and additional carriers beyond the band edges are generated. Theoretically, this carrier multiplication considerably enhances the photo-carrier density, promisingly improving the short-circuit current and resulting in PCE toward beyond theoretical SQ limit ~44% [35, 36]. However, the second mechanism known as hot carrier extraction primarily focused on the capturing of photo-excited carriers immediately before their excess energy is lost in the form of heat. The extraction of these photo-excited carrier at energies above the band gap of the semiconductor possibly delivered the increased open-circuit voltage with PCE limit up to 67% [35, 36].

4.3 Ultimate and Quantum (Internal, External) Efficiency

The photoelectrical devices mainly operate under the photoelectric effect and include the avalanche photodiode, charge coupled devices, photo-detectors, imaging sensors, light emitting diode, and solar cell devices. These photoelectrical devices convert the incident photon or light energy into the respective output such as electrical energy in case of solar cell device. The ability of converting the light by these photoelectric devices is primarily characterized by the parameter known as quantum efficiency. The quantum efficiency (QE) widely known as incident photon to charge carrier efficiency (IPCE) defined as number of charge carriers that are generated in the device with respect to the number of incident photon of light and mainly expressed as a function of incident photon wavelength and energy. The QE measurements offer the insights of the effectiveness of the PV device which led to conversion of incident photon into electron at each or a specific wavelength [37].

The capability of PV devices in converting the light energy into the electrical energy established and supported by device's spectral response (SR). The index SR is an essential parameter for evaluating and measuring the light-to-electrical conversion capability of a device [38]. It reflects the ability of generating the photocurrent by the solar cell in amperes under monochromatic illumination of the incident wavelength with the spectral irradiance of the same incident wavelength on the device and expressed as

$$\mathrm{SR}(\lambda) = \frac{I(\lambda)}{P(\lambda)} \tag{4.1}$$

where incident photon energy for each wavelength in watts is given by $P(\lambda)$ and defined by number of photons per unit time (photon/sec). However, the photo-generated current in the solar cell device in amperes (A) is given by $I(\lambda)$ and is given by the number of electrons per unit time (electron/sec).

The QE described as the ratio of number of incident photon to the number of charge carriers produced in the device and mainly expressed in percentage. The QE measurements are directly interlinked with the spectral response of the devices. If the ν is the frequency of incident photon, h is Plank's constant, λ is wavelength of the incident photon, and n is the number of electrons, then spectral response of the device is given by

$$SR(\lambda) = \frac{I(\lambda)}{P(\lambda)} = \frac{n}{h\nu} \times QE(\lambda) = \frac{\lambda}{1240} \times QE(\lambda) \tag{4.2}$$

$$QE(\lambda) = 100 \times \frac{1240 \times SR(\lambda)}{\lambda}(\%) \tag{4.3}$$

The QE based on the treatment of photon reflected and transmitted from the solar cells can be categorized into external and internal quantum efficiencies. The external QE (EQE) takes into consideration the total number of photons including the reflected and transmitted photon incident on the surface of the device. Thus, EQE widely known as QE or IPCE is described as the ratio of number of charge carriers generated to the total number of incident photon. However, the IQE takes into account the actual number of photon that takes part in generating the charge carriers in the device, i.e., excluding the reflected and transmitted photon. Thus, the IQE is defined as the ratio of number of charge carriers generated to the number of actual photons that are mainly responsible for the generation of charge carriers [37–41]. The IQE calculation facilitates the calculation of photon–electron conversion efficiency. If the reflectivity (R) and transmittance (T) of the surface of solar cell device taken into account then the relationship between the EQE and IQE can be expressed as

$$IQE(\lambda) = \frac{EQE(\lambda)}{(1 - R - T)} \times 100(\%) \tag{4.4}$$

The QE/spectral response of the PV device provides the efficiency of conversion of photon to charge carriers at different wavelengths of the incident light spectrum. There are various factors, including the light-absorbing materials' structure, manufacturing process, device structure itself, etc., which affect the conversion efficiency of the solar cell device. Thus, it is found to have different efficiencies of the solar cell at different wavelengths. However, if solar cell absorbed all incident photons with certain wavelength and the charge carriers are generated then the QE at that wavelength is unity. However, the QE for incident photon with energy below the band gap of the absorbing material is found to be zero. Figure 4.3 shows the QE pattern including the QE response of an ideal solar cell device with respect to the wavelength of the visible light spectrum.

4.4 Green ME-MF Materials

The ME materials are primarily governed by the ME effect in which coupling between magnetic and electric properties offers unique window of opportunities for less invasive integration into the devices on account of developing electric charges without any connection of wire. The ME effect in the ME materials described as the change

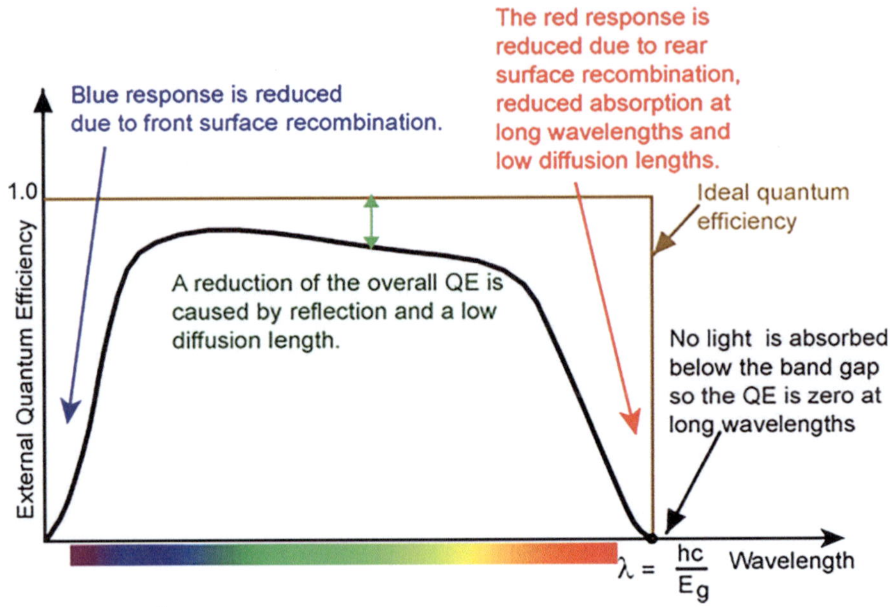

Fig. 4.3 QE illustration of a silicon solar cell. Adapted with permission from Ref. [39]

occurred in the electric polarization (P) of the material when the magnetic field (H) is applied (direct effect) [42]. However, converse ME effect is well known as when the change occurred in the magnetization (M) of the materials under the influence of applied electric filed (E). The MF materials that exhibit simultaneously both the FE and ferromagnetic ordering displayed the remarkable ME effects. However, the material which develops the electric polarization under the influence of applied magnetic field also displayed the ME effect. These ME materials are found of great interest and capable of developing the new-age multifunctional devices including high-speed memories, ME sensors, PV devices, etc. [15, 42–50].

The ME composites artificially created ME effect at room temperature that is regulated either by the polarization with respect to an applied magnetic field or to an extent induce magnetization in response to externally applied electric filed [51]. Moreover, in the ME composites coupling between the magnetic and electric properties offers a unique advantage for developing the charges without any electrical connection for the purpose of their application in the various fields. However, the ME composites with ferromagnetic (FM) and piezoelectric (PE) phases widely investigated as much larger ME effects are displayed by them. The ME effects in such materials mainly attributed to the mechanical coupling of the FM and PE phases at the interfaces which has resulted due to magnetostriction of FM phase and piezoelectricity in PE phase. The FM phase found to deformed when external magnetic field (H) applied to such ME composites due to magnetostriction and this deformation transferred to the present PE phase which further result in the generation of an electric field (E) on account of piezoelectric effect (direct ME effect). However, when such composites are subjected to an external field (E), the deformation in the PE phase arises due to converse piezoelectric effect and then this deformation is transferred to the present FM phase and on account of inverse magnetostriction the magnetic field (B) appeared to change (indirect ME effect) [52, 53].

The ME effect in the MF materials with its inherent ability of mutually altering the magnetic and electric field widely investigated for the development of new frontiers and advancement of various science and technological fields. The ME effect primarily attributes for the development of permanent and low-frequency sensors at room temperature for transforming the magnetic field with high degree of sensitivity, and in the making of solid-state inductors and transformers mainly tuned by the electric field, energy harvesters, converting AC magnetic field into DC voltages and used in the compact low-frequency antennas, electrically tunable resonators, and microwave filters and data storage elements altered by an electric field [54–59]. One of the most promising fields is PV field where ME materials could be employed for the generation of green energy.

4.5 ME-MF Materials for Green Energy Applications

The sun is the natural source of energy for living and non-living elements of our planet and radiates energy in all directions in the entire range of electromagnetic spectrum. Approximately, 45% of the solar spectrum constitutes the solar radiation ranging from 400 to 700 nm known as visible spectrum and human eye is found to be sensitive to this narrow range of solar spectrum. The solar radiation entered into the earth atmosphere if harvested adequately found to be phenomenal source of energy and has the potential to power the planet earth. Apart from the fossil fuel reserve, solar radiation is found to be the largest global reserve by source among the most important renewable energy sources including the wind energy, hydro, geothermal, and biomass energy. The sun continuously supplying solar radiation in great amount and green energy by far is the most suitable source for the renewable energy harvesting. In order to harvest the solar radiation, it is the dire need for developing more efficient and less expensive PV technologies so that energy requirements of the planet could be solved regardless of increase in the annual usage due to increase in the population growth.

Two basic processes mainly involved in the PV effect, one is electron–hole pair generation as the electrical-charge carriers and other is the formation of net current after the separation of electrons and holes in a particular direction. The standard solar cell based on *p–n* semiconductor junction produced the electric current resulted from the separation of electron and holes when the incident photon excited them across from the valence to the conduction band gap. The value of the band gap of the semiconductor and the value of the energy barrier formed at depletion layer at the *p–n* junction somehow limited to 1 V [60]. Thus, the large energy photon with respect to energy band gap of the semiconductor can produce the electric current. Thus, the small band gap semiconductor absorbs more incident photon and results in the large current flow but delivers low voltage. However, the large band gap semiconductor delivers the high value voltage with limited current value as the large number of incident photon pass through the semiconductor without absorption. In light of this, a small fraction of solar radiation of ~2 eV visible photons can only be harvested with the present PV technology based on semiconductor [61].

Moreover, the bulk PV effect in the FE materials have emerged as the solution to the solar power problem which arises in the *p–n* junction-based semiconductor solar cell. When the FE material exposed to the incident photon with energies larger than the band gap of the given FE material appeared to generate electron and holes as the photons are absorbed within the bulk of FE crystal. The electric filed associated with the internal polarization (P) of the FE material causes the separation of the generated charge carriers toward the respective electrode. Thus, the PV effect in the FE materials fundamentally is a bulk effect and considerably different from the conventional interfacial effect of the *p–n* junction semiconductor in which charge is separated due to electric field present at the depleted space at *p–n* junction interface. The remnant polarization (P_r) of the FE material is primarily manifested in the form of internal electric field (E_i) that is present in the whole bulk region of the given FE instead of limited to the thin interface. The PV effect due to interfacial region of the semiconductor is mainly influenced by the diffusion and the energy barrier and thus restricted the open-circuit voltage in *p–n* junction semiconductor up to ~1 V. However, in the bulk PV effect, the diffusion of charge carriers and energy barrier formation do not restrict charge transport, thus leading to offer the high open-circuit photo-voltage. Figure 4.4 shows the schematic bulk PV effect in the MF materials.

However, the MF materials displayed more than one ferroic order parameters expected to replace the FE material for the absorption of photon of wider solar spectrum as the additional degree of freedom with magnetic order parameter or the intrinsic strain coupling definitely aids in lowering the band gap of the MF structure. The whole solar spectrum possibly gets harnessed in the process of photon to electrical energy conversion in the bulk PV cell with band gap less than 1 eV. The ME materials among the MFs are suggested to be deployed as at photoactive material for the development of new PV device. There are various recent reports available in which ME materials are successfully employed for absorbing the incident photon. The Cr-doped $BiFeO_3$ (Bi_2CrFeO_6) ME thin films with measured band gap around 1.4–2 eV reported the large PV effect with response to whole incident solar radiation. In this two BFCO films A and B were directly grown at different temperatures by pulsed laser deposition on niobium-doped $SrTiO_3$ (STO:Nb) substrate with indium-doped tin oxide electrode (ITO) on top. The schematic solar cell device is shown in the top-left inset of Fig. 4.5a. The open-circuit photo-voltage of 0.6 V was observed under the illumination of red light with 1.5 mW/cm^2 intensity with 6% PCE when the thickness of the ME film was 125 nm [62]. Subsequently, electric polarization reversal results in the change in the photocurrent, however the polarization reversal causes no change in the dark current. Figure 4.5a shows the *J–V* measurement curve under the dark and light illumination for both the films with the directional leakage at positive voltage which reflects the diode-like behavior and suggesting the PV effect

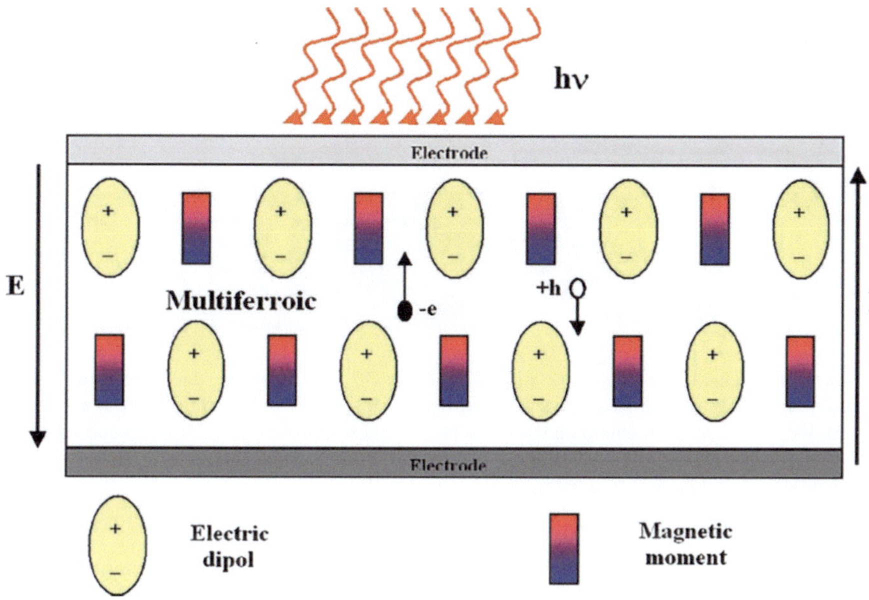

Fig. 4.4 Schematic representation of the bulk PV effect in the MF materials. Adapted with permission from Ref. [6]

in both the Cr-doped $BiFeO_3$ ME thin epitaxial films [62]. The FE properties of the compressively strained in-plane epitaxial grown BFCO films were studied by using the piezoresponse force microscopy to further explore the origin of the enhanced PV effect in the grown film. The PFM images are shown in Fig. 4.5b–f of as-grown film B (Fig. 4.5b) and A (Fig. 4.5e) and after switching the FE domain of the film B (Fig. 4.5c) and film A (Fig. 4.5f) with their respective hysteresis loops of films B and A (Fig. d and g). The prominent polarization imprint is found in film A as shown in Fig. 4.5e with FE polarization at zero external voltage established the existence of intense internal electric filed caused by the separation of charges that are generated due to photoelectric effect and result in large photocurrent.

In another study, optical and PV properties along with structural, magnetic, and electrical one were extensively explored in Pr-doped $BiFeO_3$ ($Bi_{1-x}Pr_xFeO_3$:$x = 0.0$, 0.05, 0.10, 0.15, 0.20) which is synthesized by sol–gel method. The FE properties of the $BiFeO_3$ found to be improved and the band gap reduced with increase in the doping of Pr in the $BiFeO_3$ [63]. The measured I–V characteristics of the pure and Pr-doped BFO at room temperature further established the outcome of the spontaneous polarization and the band gap on the PV response. The I–V characteristics of the pure BFO sample are observed to be symmetrically aligned with the polarity change and magnitude of the current is found to be increased with the external applied bias as shown in Fig. 4.6a, b. In order to investigate the photoconductivity of the prepared samples, I–V characteristics of the respective samples were studied under the dark and illumination conditions as shown in the Fig. 4.6a, b, respectively. The

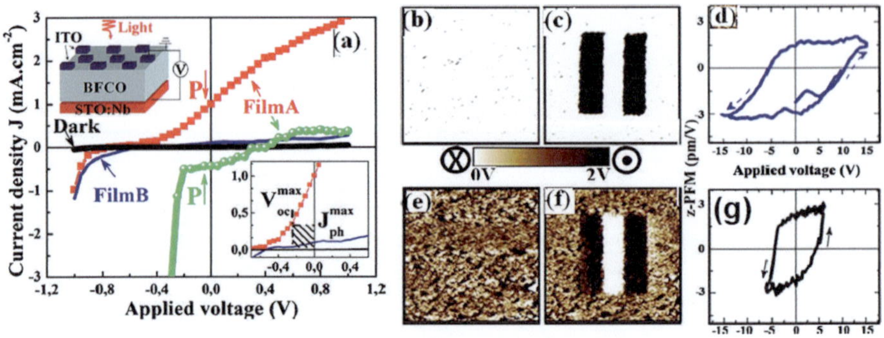

Fig. 4.5 **a** J–V characteristics of the epitaxial grown BFCO film under dark and illumination condition with schematic device illustration on top-left inset, the bottom-right inset is a zoomed view for the values taken to calculate the efficiency. **b** and **e** PFM images of films B and A. **c** and **f** PFM images after switching the FE domain of films B and A. **d** and **g** the hysteresis loop of the as-grown films B and A. Adapted with permission from Ref. [62]

photon of wavelength 405 nm with 5 mW maximum power from the laser was used for the purpose of producing current that is induced by the photon excitation. The value of photocurrent under the illumination condition was found to be modified. The excitation of charge carriers across the Schottky barrier formed between ITO and BFO causes the shift in the conductivity under the illumination condition. The change in the conductivity appears to be quite high when the external field is applied with respect to I–V response under dark condition up to Pr = 0.15. The band gap modification and improved FE properties with doping of Pr in BFO might offer the background for increasing the charge carriers induced by the incident photon and their separation due to present electric filed. As a result of this study, Pr-doped BFO samples were found to have improved conductivity under the illumination condition which is the essential requirement for employing the BFO for the PV applications [63].

The newly found MF material $InFeO_3$ possesses the antiferromagnetism and high FE polarization along with structure similar to $BiFeO_3$ [64]. To comprehensively investigate the ME properties, photocatalysis, and PV performances, theoretically the doping of 3d transition metal in the $InFeO_3$ carried out to form In_2FeX (X = V, Cr, Mn, Co, and Ni)O_6. $InFeO_3$ and $BiFeO_3$ share the same structures and ME properties. The band gap of the photoactive materials plays the crucial role in affecting the PV conversion efficiency of that material. The band gap of In_2FeVO_6 is designed ~2.0 eV and the presence of adequate FE polarization makes it suitable for the carriers' separation. From the stability and band gap analysis point of view, In_2FeVO_6 can be established as potential photoactive material in the In_2FeX(X = V, Cr, Mn, Fe, Co, and Ni) O_6. There is a model known as spectroscopic limited maximum efficiency (SLME) based on band gap of the light-absorbing material and the absorption coefficient, thickness, and non-equilibrium radiation proposed to calculate the PCE of the ideal light-absorbing materials [65].

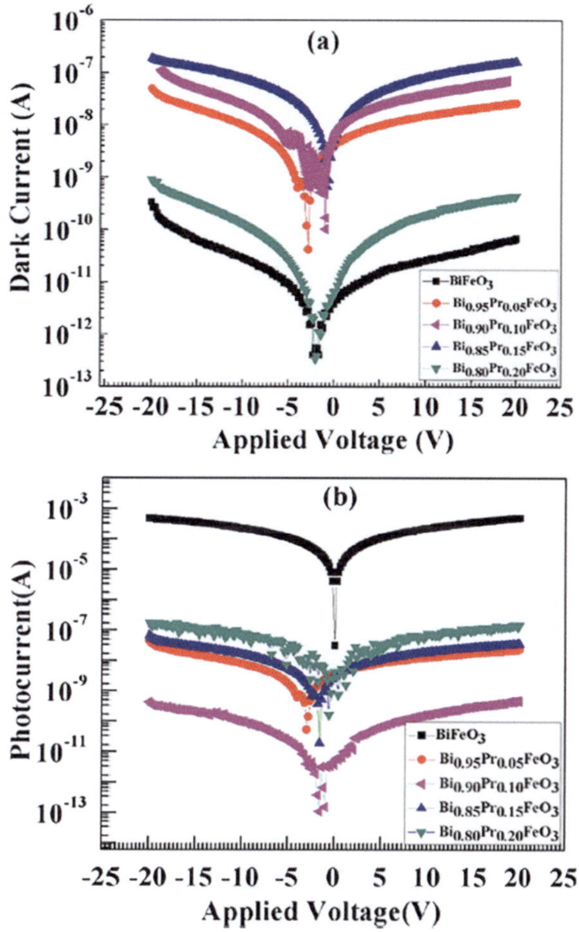

Fig. 4.6 PV responses of the Pr-free and Pr-doped $BiFeO_3$ under, **a** dark and **b** illumination condition. Adapted with permission from Ref. [63]

The absorption coefficients of In_2FeVO_6 were calculated under the intrinsic and compressive and tensile strain condition and found to be varied with incident photon wavelength with higher value 10^5 cm^{-1} as shown in Fig. 4.7a [64]. The high value of absorption coefficients shows that the In_2FeVO_6 could absorb the sufficient amount of solar radiation from visible spectrum. Figure 4.7b shows the maximum PCE of the In_2FeVO_6 under the strained condition, viz., compressive and tensile at different thicknesses of the light absorber materials calculated by the SLME model. The strain was introduced by altering the respective lattice parameters. The calculated result categorically reflects that maximum PCE ~20% can be reached when In_2FeVO_6 is employed as photoactive material in the solar cell devices [64]. The different parameters of In_2FeVO_6-based solar cell along with other solar cells like short-circuit current and open-circuit voltage were calculated experimentally and theoretically listed in Table 4.1 [41, 64].

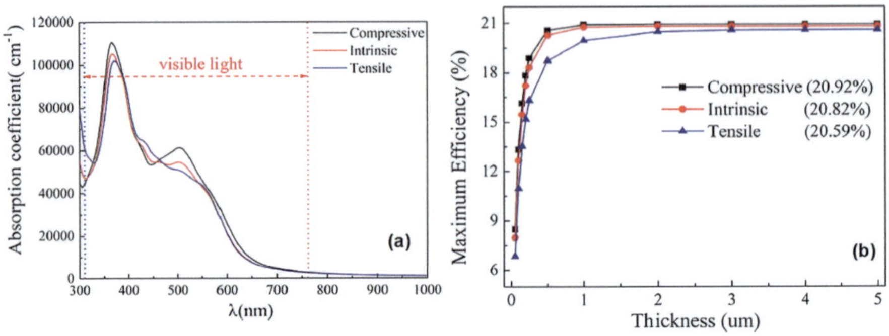

Fig. 4.7 Calculated, **a** absorption coefficient and **b** maximum PCE of with respect to incident photon wavelength and thickness of the absorbing material, respectively. Adapted with permission from Ref. [64]

Table 4.1 PV performance parameters of various solar cell devices based on different photoactive materials obtained theoretically and experimentally where C and T represent the compressive and tensile strained samples

Materials	J_{sc} (mA/cm^2)	V_{oc} (V)	PCE (%)	References
C-In_2FeVO_6 (theoretical)	14.26	1.62	20.92	Adapted with permission from Ref. [64]
In_2FeVO_6 (theoretical)	14.41	1.60	20.82	
T-In_2FeVO_6 (theoretical)	17.75	1.32	20.59	
Si (theoretical)	39.71	0.79	27.08	
Si (experimental)	42.65	0.738	26.7	
Bi_2FeCrO_6 (experimental)	20.6	0.84	8.1	
Er-doped BFO (theoretical)	22.18	0.79	13.49	[41]

The photoelectric conversion efficiency of the In_2FeVO_6-based solar cell managed to be at around 20% due to outcome of the strain. However, the PV efficiency remains unaltered despite the tensile strain causes the increase in the short-circuit current and slightly reduces the open-circuit voltage. As the most MF materials found to be stable under the high-pressure condition, the high PCE (~20%) of the In_2FeVO_6-based solar cell is maintained under the 5 Gpa calculated by the SLME model. The Bi_2FeCrO_6-based thin film solar cell with multilayer structure experimentally provides 0.84 V open-circuit voltage and short-circuit current ~20.6 mA/cm^2. Here, it is firmly established that In_2FeVO_6-based solar cell managed to provide the higher open-circuit voltage in comparison to the Bi_2FeCrO_6-based thin film solar cell. Table 4.1 provides the various parameters of the traditional silicon solar materials and compared with the In_2FeVO_6-based solar cell which shows comparatively higher open-circuit voltage and smaller short-circuit current for In_2FeVO_6-based solar cell. The various PV performance parameters of the In_2FeVO_6 appeared to be saturated at the thickness about 2 μm and exhibit the FE characteristic. The FE polarization values of In_2FeVO_6 calculated theoretically under the 2% compressive and tensile strained conditions which were found to be 91.23 μc/cm^2 and 81.67 μc/cm^2, respectively. These values categorically reflect the strength of high FE polarization which further make In_2FeVO_6 a suitable and potential candidate of high power conversion material for the PV application [41, 64, 65].

A recent study by Kumar et al. performed theoretically via SCAPS-1D software investigated for the very first time Er-doped $BiFeO_3$ MF as the photoactive material and designed PSC devices [41]. The modified band gap of the Er-doped BFO with respective composition of Er (0.04, 0.08, 0.12%) was employed to computationally design the four devices with cell configuration FTO/ETL/Absorber/Spiro-OMeTAD/Au. This simulation work was performed under the illumination of AM 1.5 G. The optimization of the designed devices was carried out with purpose of obtaining the superior PCE. The effect of thickness and defect density of the absorber layer and shunt and series resistance was studied thoroughly. The effect of device operating temperature and the effect of various electron transport layer made of different materials were also investigated. The final optimized purposed PSC device exhibits the superior PV performance parameters as shown in Table 4.1 including PCE ~13.49%. These results unconditionally state that the Er-doped BFO could be the potential contender for high-efficiency PV applications [41].

4.6 Opportunities and Future Challenges

The MF materials possess simultaneously alterable electric/dielectric, magnetic, mechanical, thermal, and optical properties and these properties made them widely interesting and yet to be explored for critical application in the area of memory devices, optoelectronic devices, sensors, and PV devices. ME materials among the MF mainly comprised ferroelectricity and magnetism together and composed of different atoms. The FE character in the ME-MF materials resulted in the existence of intense internal electric field that facilitates the abnormal bulk PV effect in materials and has excited the interest of PV research and scientific community worldwide. The presence of FE character in the ME-MF materials provides an opportunity to get ultra-high open-circuit voltage and smoothly enabled charge separation unlike conventional PV solar cell which offer limited open-circuit voltage. However, the PCE delivered by the ME-MF-based PV devices is yet to be improved though the research for these purposes is in the very promising stage making them to compete with their counterparts available in the PV market [66–68].

The multiple parameters including the ferroelectricity, ferromagnetism, and ferroelasticity can couple to each other and provide a fascinating cross-coupling effect such as ME effect that to a great extent intensified tremendously the versatility of the ME-MFs for novel device application. The mutually dependent multiple functionalities in ME-MF material mainly become a deciding factor for employing them in different applications. The strictly observed requirement for the various application purposes is presence of strong ME coupling effect rather room temperature, simple, easy, and cost-effective manufacturing of materials and devices with non-toxic and environment-friendly chemical elements and compounds. The ME-MF material's structure required to be suitable for large-scale manufacturing so that it could be integrated at chip/wafer level. ME-MF structure further required to be appropriate and desirable without losing functionalities for the fabrication at different levels including micro- and nanoscale size. The experimental retaining of magnetoelectric effect at nano-level has been reported which further opens up a window of opportunities for employing these materials with multiple functionalities in nano-technological field [66–72].

In the light of global energy demand, the generation of electricity from the renewable energy source, i.e., the non-toxic and environment friendly is mainly required for the sustainable development and progress. ME-MF materials provide an opportunity to be employed as a photoactive material under the bulk PV effect. ME-MF materials with their efficient FE polarization separate the charge carriers and facilitate to generate the above band gap photo-voltage. However, ME-MF materials were found with high band gap. So far, the biggest challenge in employing ME-MF materials as a light-absorbing material is to reduce their energy band in the range of visible spectrum while simultaneously holding back their FE character. The development of multilayer structures and the doping with suitable material were found to be the effective methods in order to enhance the photoelectric/power conversion efficiency of the ME-MF materials that allows retaining the sufficient strength of the FE polarization and reduces the band gap toward the visible range [73, 74].

4.7 Concluding Remarks

In the light of the above discussion, it is attempted to discuss the various possibilities of ME-MF materials for the employment as a photoactive material for the conversion of photon energy to the electrical energy. The SQ limit for the single-junction semiconductor solar cell was discussed. The theoretical limit of maximum PCE of a single-junction solar cell and the causes of this limit were also elaborated. The possible way of crossing the upper limit of the efficiency was briefly discussed. The QE includes the internal and external both as they played the essential role in deciding and identifying the incident photon participating in the conversion processes of the PV devices. ME-MF materials and the mechanism of ME effect were also discussed. ME-MF materials which demonstrate the bulk photovoltaic effect are found to be the new-age materials for photoactive material in the PV field if their band gap is reduced while retaining the FE character in them intact.

References

1. Goetzberger A, Hebling C (2000) Photovoltaic materials, past, present, future. Sol Energy Mater Sol C 62(1–19):1. https://doi.org/10.1016/S0927-0248(99)00131-2
2. Yuan Y, Xiao Z, Yang B, Huang J (2014) Arising applications of ferroelectric materials in photovoltaic devices. J Mater Chem A 2:6027. https://doi.org/10.1039/c3ta14188h
3. Lewis NS (2007) Toward cost-effective solar energy use. Science 5813:798–801. https://doi.org/10.1126/science.1137014
4. Trupke T, Shalav A, Richards BS, Würfel P, Green MA (2006) Green, Efficiency enhancement of solar cells by luminescent up-conversion of sunlight. Sol Energy Mater Sol C 90:3327. https://doi.org/10.1016/j.solmat.2005.09.021
5. Jalaja MA, Dutta D (2015) Ferroelectric and multiferroic for next generation photovoltaic. Adv Mater Lett 7:568–584. https://doi.org/10.5185/amlett.2015.5878
6. Vopson MM (2014) Fundamentals of multiferroic materials and their possible applications. Vopson 40:223–250. https://doi.org/10.1080/10408436.2014.992584
7. Chynoweth AG (1953) Surface space-charge layers in barium titanate. Phys Rev 102:705. https://doi.org/10.1103/PhysRev.102.705
8. Brody PS (1973) Large polarization-dependent photovoltages in ceramic $BaTiO_3$+ 5wt% $CaTiO_3$. Solid State Commun 12:673. https://doi.org/10.1016/0038-1098(73)90310-4
9. Chanussot G (1978) Physical models for the photoferroelectric phenomena, Ferroelectrics 20:37–50. https://doi.org/10.1080/00150197808239428
10. Yang SY, Seidel J, Byrnes SJ, Shafer P, Yang CH, Rossell MD, Yu P, Chu YH, Scott JF, Ager JW, Martin LW, Ramesh R (2010) Above-bandgap voltages from ferroelectric photovoltaic devices. Nat Nanotechnol 43:143–147. https://doi.org/10.1038/nnano.2009.451
11. Koch TH, Munser R, Ruppel W, Wurfel P (1975) Bulk photovoltaic effect in $BaTiO_3$. Solid State Commun 17:847–850. https://doi.org/10.1016/0038-1098(75)90735-8
12. Glass M, Von der Linde D, Negran TJ (1974) High_voltage bulk photovoltaic effect and the photorefractive process in $LiNbO_3$. Appl Phys Lett 4(25):233–235. https://doi.org/10.1063/1.1655453
13. Landsberg T (1975) An introduction to the theory of photovoltaic cells. Solid State Electron 18:1043. https://doi.org/10.1016/0038-1101(75)90165-3
14. Spaldin NA, Cheong S-W, Ramesh R (2010) Multiferroics: past, present, and future. Phys Today 63:38–43. https://doi.org/10.1063/1.3502547
15. Eerenstein W, Mathur ND, Scott JF (2006) Multiferroic and magnetoelectric materials. Nature 442:759–765. https://doi.org/10.1038/nature05023
16. Ramesh R (2022) Materials for a sustainable microelectronics future: electric field control of magnetism with multiferroics. J Indian Inst Sci 102:489–511. https://doi.org/10.1007/s41745-021-00277-7
17. Sun B, Zhou G, Sun L, Zhao H, Chen Y, Yang F, Zhaoab Y, Song Q (2021) ABO_3 multiferroic perovskite materials for memristive memory and neuromorphic computing. Nanoscale Horiz 6:939–970. https://doi.org/10.1039/d1nh00292a
18. Tu C, Dong C, Chu Z, Chen H, Liang X, Sun NX (2018) Apassive isolator realized by magnetoelectric laminate composites. Appl Phys Lett 113:262904. https://doi.org/10.1063/1.5080346
19. Zaeimbashi M, Lin H, Dong C, Liang X, Nasrollahpour M, Chen H, Sun N, Matyushov A, He Y, Wang X et al (2019) Ultra-compact dual-band smart NEMS magnetoelectric antennas for simultaneous wireless energy harvesting and magnetic field sensing. IEEE J Electromagn RF Microw Med Biol 3:206. https://doi.org/10.1038/s41467-021-23256-z
20. Tu C, Chu Z-Q, Spetzler B, Hayes P, Dong C-Z, Liang X-F, Chen H-H, He Y-F, Wei Y-Y, Lisenkov I, Lin H, Lin Y-H, McCord J, Faupel F, Quandt E, Sun N-X (2019) Mechanical-resonance-enhanced thin-film magnetoelectric heterostructures for magnetometers, mechanical antennas, tunable RF inductors, and filters. Materials 14:2259. https://doi.org/10.3390/ma12142259

21. Liang X, Dong C, Chen H, Wang J, Wei Y, Zaeimbashi M, He Y, Matyushov A, Sun C, Sun N (2020) A review of thin-film magnetoelastic materials for magnetoelectric applications. Sensors 20:1532. https://doi.org/10.3390/s20051532
22. Dong S, Zhai J, Li JF, Viehland D, Priya S (2008) Multimodal system for harvesting magnetic and mechanical energy. Appl Phys Lett 93:103511. https://doi.org/10.1063/1.2982099
23. Onuta T-D, Wang Y, Long CJ, Takeuchi I (2011) Dynamic state switching in nonlinear multiferroic cantilevers. Appl Phys Lett 101:203506. https://doi.org/10.1063/1.4738991
24. Kosub T, Kopte M, Huhne R, Appel P, Shields B, Maletinsky P, Hubner R, Liedke MO, Fassbender J, Schmidt OG (2017) Purely antiferromagnetic magnetoelectric random access memory. Nat Commun 8:13985. https://doi.org/10.1038/ncomms13985
25. Bibes M, Barthelemy A (2008) Multiferroics: towards a magnetoelectric memory. Nat Mater 7:425. https://doi.org/10.1038/nmat2189
26. Li X, Lee A, Razavi SA, Wu H, Wang KL (2018) Voltage-controlled magnetoelectric memory and logic devices. MRS Bull 12:970. https://doi.org/10.1557/mrs.2018.298
27. Grezes C, Ebrahimi F, Alzate JG, Cai X, Katine JA, Langer J, Ocker B, Khalili Amiri P, Wang KL (2016) Ultra-low switching energy and scaling in electric-field-controlled nanoscale magnetic tunnel junctions with high resistance-area product. Appl Phys Lett 108:012403. https://doi.org/10.1063/1.4939446
28. Lee J-H, Murugavel P, Ryu H, Lee D, Jo JY, Kim JW, Kim HJ, Kim KH, Jo Y, Jung M-H, Oh YH, Kim Y-W, Yoon J-G, Chung J-S, Noh TW (2006) Epitaxial stabilization of new multiferoic hexagonalphase of $TbMnO_3$ thin films. Adv Mater 18:3125–3129. https://doi.org/10.1038/ncomms13985
29. Park T-J, Papaefthymiou GC, Viescas AJ, Moodenbaugh AR, Wong SS (2007) Size-dependent magnetic properties of single-crystalline multiferroic $BiFeO_3$ nanoparticles. Nano Lett 7:766–772. https://doi.org/10.1021/nl063039w
30. Shockley W, Queisser HJ (1961) Detailed balance limit of efficiency of p-n junction solar cells. J Appl Phys 32:510. https://doi.org/10.1063/1.1736034
31. https://en.wikipedia.org/wiki/Shockley%E2%80%93Queisser_limit
32. Zanatta AR (2022) The Shockley–Queisser limit and the conversion efficiency of silicon-based solar cells. Results Opt 9:100320. https://doi.org/10.1016/j.rio.2022.100320
33. Kirk AP, Cardwell DW (2017) Reconsidering the Shockley–Queisser limit of a ferroelectric insulator device.Nat Photon 11:329. https://doi.org/10.1038/nphoton.2017.84
34. Wang K, Zheng L, Hou Y, Nozariasbmarz A, Poudel B, Yoon J, Ye T, Yang D, Pogrebnyakov AV, Gopalan V, Priya S (2022) Overcoming Shockley-Queisser limit using halide perovskite platform? Joule 6:756–771. https://doi.org/10.1016/j.joule.2022.01.009
35. Rühle S (2016) Tabulated values of the Shockley–Queisser limit for single junction solar cells. Solar Energy 130:139–147. https://doi.org/10.1016/j.solener.2016.02.015
36. Lee YH (2024) Beyond the Shockley-Queisser limit: exploring new frontiers in solar energy harvest. Science 383:4308. https://doi.org/10.1126/science.ado4308
37. Karmalawi AM, Rayan DA, Rashad MM (2020) Establishment and evaluation of photovoltaic quantum efficiency system at central metallurgical research and development institute. Optik—Int J Light Electron Opt 217:164931. https://doi.org/10.1016/j.ijleo.2020.164931
38. Ananda W (2017) External quantum efficiency measurement of solar cell. In: Symposium on Electrical and Computer Engineering, pp 450–456. https://doi.org/10.1109/QIR.2017.8168528
39. https://www.pveducation.org/pvcdrom/solar-cell-operation/quantum-efficiency
40. Semonin OE, Luther JM, Choi S, Chen HY, Gao J, Nozik AJ, Beard MC (2011) Peak external photocurrent quantum efficiency exceeding 100% via MEG in a quantum dot solar cell. Science 6062:1530–1533. https://doi.org/10.1126/science.1209845
41. Kumar M, Pundir SK, Singh DV, Kumar M (2023) Effect on green energy conversion and stability with 'Er' modification in multiferroic based perovskite solar cell devices. Mater Today Commun 107841. https://doi.org/10.1016/j.mtcomm.2023.107841
42. Curie P (1894) Sur la symetrie dans les phenomenes physiques, symetrie d'un champ electrique et d'un champ magnetique. J Phys Theor Appl 3:393–415. https://doi.org/10.1051/jphystap:018940030039300

43. Chen H, Liang X, Dong C, He Y, Sun N, Zaeimbashi M, He Y, Gao Y, Parimi PV, Lin H, Sun N-X (2020) Ultra-compact mechanical antennas. Appl Phys Lett 117:170501. https://doi.org/10.1063/5.0025362
44. Kemp MA, Franzi M, Haase A, Jongewaard E, Whittaker MT, Kirkpatrick M, Sparr R (2019) A high Q piezoelectric resonator as a portable VLF transmitter. Nat Commun 10:1715. https://doi.org/10.1038/s41467-019-09680-2
45. Pereira N, Lima AC, Correia V, Perinka N, Lanceros-Mendez S, Martins P (2020) Magnetic proximity sensor based on magnetoelectric composites and printed coils. Materials 13:1729. https://doi.org/10.3390/ma13071729
46. Vidal JV, Turutin AV, Kubasov IV, Kislyuk AM, Kiselev DA, Malinkovich MD, Parkhomenko YN, Kobeleva SP, Sobolev NA, Kholkin AL (2020) Dual vibration and magnetic energy harvesting with bidomain $LiNbO_3$-based composite. IEEE Trans Ultrason Ferroelectr Freq Control 67:1219–1229. https://doi.org/10.1109/TUFFC.2020.2967842
47. Schmid H (2008) Some symmetry aspects of ferroics and single phase multiferroics. J Phys Condens Matter 20:434201. https://doi.org/10.1088/0953-8984/20/43/43420
48. Jiang Q, Zhong CG (2002) Magnetoelectric properties in A-type antiferromagnet in an electric field. J Mag Mag Mater 247:6–14
49. Folen VJ, Rado GT, Stalder EW (1961) Anisotropy of the magnetoelectric effect in Cr_2O_3. Phys Rev Lett 6:607–608. https://doi.org/10.1103/PhysRevLett.6.607
50. Popov YF, Kadomtseva AM, Vorob'ev GP, Zvezdin AK (1994) Discovery of the linear magnetoelectric effect in magnetic ferroelectric $BiFeO_3$ in a strong magnetic field. Ferroelectrics 162:135–140. https://doi.org/10.1080/00150199408245098
51. Zheng T, Yue Z, Wallace GG, Du Y, Martins P, Lanceros-Mendez S, Higgins MJ (2017) Local probing of magnetoelectric properties of PVDF/Fe_3O_4 electrospun nanofibers by piezoresponse force microscopy. Nanotechnology 28:65707. https://doi.org/10.1088/1361-6528/aa5217
52. Van Den Boomgaard J, Terrell DR, Born RAJ, Giller HFJI (1974) An in situ grown eutectic magnetoelectric composite material. J Mater Sci 9:1705–1709. https://doi.org/10.1007/BF00540770
53. Van Suchtelen J (1972) Product properties: a new application of composite materials. Philips Res Rep 127:28–37
54. Palneedi H, Annapureddy V, Priya S, Ryu J (2016) Status and perspectives of multiferroic magnetoelectric composite materials and applications. Actuators 5:9. https://doi.org/10.339s0/act5010009
55. Viehland D, Wuttig M, McCord J, Quandt E (2018) Magnetoelectric magnetic field sensors. MRS Bull 43:834–840. https://doi.org/10.1557/mrs.2018.261
56. Saveliev D, Chashin D, Fetisov L, Shamonin M, Fetisov Y (2020) Ceramicheterostructure-based magnetoelectric voltage transformer with an adjustable transformation ratio. Materials 13:3981. https://doi.org/10.3390/ma13183981
57. Narita F, Fox M (2017) A review of piezoelectric, magnetostrictive, and magnetoelectric materials and device technologies for energy harvesting applications. Adv Eng Mater 20:1700743. https://doi.org/10.1002/adem.201700743
58. Srinivasan G, Fetisov YK (2005) Microwave magnetoelectric effects and signal processing devices. Integr Ferroelectr 83:89–98. https://doi.org/10.1080/10584580600949105
59. Kleemann W (2017) Multiferroic and magnetoelectric nanocomposites for data processing. J Phys D Appl Phys 50:223001. https://doi.org/10.1088/1361-6463/aa6c04
60. Cummerow RL (1954) Photovoltaic Effect in $p{-}n$ Junctions. Phys Rev 95:16. https://doi.org/10.1103/PhysRev.95.16
61. López N, Reichertz LA, Yu KM, Campman K, Walukiewicz W (2011) Engineering the electronic band structure for multiband solar cells. Phys Rev Lett 106:028701. https://doi.org/10.1103/PhysRevLett.106.028701
62. Nechache R, Harnagea C, Licoccia S, Traversa E, Ruediger A, Pignolet A, Rosei F (2011) Photovoltaic properties of Bi_2FeCrO_6 epitaxial thin films. App Phys Lett 98:202902. https://doi.org/10.1063/1.3590270

63. Artia S, Kumara P, Kumara R, Walia V (2019) Vermaa, Improved ferroelectric, magnetic and photovoltaic properties of Pr doped multiferroic bismuth ferrites for photovoltaic application. Res Phys 14:102403. https://doi.org/10.1016/j.rinp.2019.102403
64. Chen X-Y, Su K-R, Ling S-W, Li C-S, Lai G-X, Luo G-P, Xu X-F, Ji H, Tang J-J (2022) Theoretical study on the multiferroic materials In2FeX ($X = V$, Cr, Mn, Co, and Ni) O_6 for high photovoltaics and photocatalysis performance. Results Phys 35:105368. https://doi.org/10.1016/j.rinp.2022.105368
65. Yu L, Zunger A (2012) Identification of potential photovoltaic absorbers based on first-principles spectroscopic screening of materials. Phys Rev Lett 108(6):068701. https://doi.org/10.1103/PhysRevLett.108.068701
66. Caruntu G, Yourdkhani A, Vopsaroiu M, Srinivasan G (2012) Nanoscale 10:3218–3227. https://doi.org/10.1039/C2NR00064D
67. Hill NA (2000) Why are there so few magnetic ferroelectrics? J Phys Chem B 104:6694–6709. https://doi.org/10.1021/jp000114x
68. Fiebig M, Lottermoser T, Meier D et al (2016) The evolution of multiferroics. Nat Rev Mater 8:1–14. https://doi.org/10.1038/natrevmats.2016.46
69. Catalan G, Scott JF (2009) Physics and applications of bismuth ferrite. Adv Mater 24:2463–2485. https://doi.org/10.1002/adma.200802849
70. Heron JT, Bosse JL, He Q, Gao Y, Trassin M, Ye L et al (2014) Deterministic switching of ferromagnetism at room temperature using an electric field. Nature 516:370–373. https://doi.org/10.1038/nature14004
71. Guzelturk B, Mei AB, Zhang L, Tan LZ, Donahue P, Singh AG et al (2020) Light-induced currents at domain walls in multiferroic $BiFeO_3$. Nano Lett 1:145–151. https://doi.org/10.1021/acs.nanolett.9b03484
72. Choi T, Lee S, Choi YJ, Kiryukhin V, Cheong S-W (2009) Switchable ferroelectric diode and photovoltaic effect in $BiFeO_3$. Science 5923:63–66. https://doi.org/10.1126/science.1168636
73. Yao Q, Liu W, Zhang R, Ma Y, Wang Y, Hu R et al (2019) Enhanced adjustable photovoltaic response in multilayer $BiFeO_3$ films. ACS Sustain Chem Eng 14:12439–12446. https://doi.org/10.1021/acssuschemeng.9b01984
74. Zhang Y, Yang Y, Dong Z et al (2020) Enhanced photocatalytic activity of Ba doped $BiFeO_3$ by turning morphologies and band gap. J Mater Sci: Mater Electron 18:15007–15012. https://doi.org/10.1007/s10854-020-04064-5

Chapter 5
Magnetoelectric Multiferroics for Spintronic Applications

5.1 Introduction

Electrons possess charge and have associated intrinsic spin. Both spin and charge were considered to in separate existence until in 1988 when Baibich reported Giant magnetoresistance (GMR). The discovery of GMR laid the foundation for the making of hard-drive read head which were introduced commercially by IBM in 1991. Thereafter, the subsequent decades saw emergence of spintronics and spintronic-based devices [1–3]. Also, in 1991, Bruno and Chappert reported existence of an oscillatory coupling or FM-AFM coupling present in between two FM layers separated by a non-magnetic metal spacer layer [4]. The interlayer exchange coupling depicted an oscillatory behavior and can be represented in terms of Fermi surface of spacer. Here, the magnetic free layer finds applications in magnetic tunnel junction (MTJ) and Multiferroic tunnel junctions (MFTJ) device. Thereafter, emerged the concept of 'spin valve' which finds applications in magnetic sensors and MRAM i.e. magnetic random access memory devices.

Spintronics, as the name suggests is 'spin transport electronics' or 'spin-based electronics', relates electron's intrinsic spin and its magnetic moment. Here, spin is an intrinsic angular momentum of an electron whereby, every electron acts as a tiny magnet and exhibits a strong magnetic moment ($\mu_B = 10^{-23}$ Am^2). Spin can be either clockwise or anticlockwise. In other words, spin can point in either upward or in downward direction. With a non-magnetic material, spins are oriented randomly, consequently the resultant spin is zero. However, with application of the magnetic field in a FM material, the spins are aligned either all up or all down. This corresponds to the binary information one (all up spins) and zero (all down spins). In quantum mechanics, the spin of a particle measured along any direction can assume specific values: $S_i = \hbar s_i$, s is principal spin quantum number, 's_i' refers to the spin of the particle projected along *i*th direction and '*i*' can possess values corresponding to all three directions i.e., *x*, *y* and *z*. Further, the spin projection quantum number i.e., s_i can have values ($-s$, $-(s + 1)$ …, $(s + 1)$, s). Conventionally, we define the

M. Kumar and S. Dabas, *Multiferroic and Magnetoelectric Materials*, Materials Horizons: From Nature to Nanomaterials, https://doi.org/10.1007/978-981-95-4391-5_5

z-component of the spin angular momentum quantum number as: $S_z = \hbar s_z$, s_z can have values (−s, $-(s+1)$ …, $(s+1)$, s). Thus, for a spin ½ particle, S_z possesses two distinct values, $+\frac{\hbar}{2}$ (spin up) and $-\frac{\hbar}{2}$ (spin down). Also, the spin can be measured from the magnetic moment i.e., $-g\mu_B s$, where, g is the Landè's g factor (for free electron $g_o = 2.0023$), μ_B is Bohr magneton ($\mu_B = \frac{e\hbar}{2m_e}$ in SI units). Spintronics relates the intrinsic spin i.e., up-spin and down-spin of electrically charged electron with its magnetic moment. Ideally, spintronics involves 'spin current' instead of 'charge current' and therefore no energy is created and/or wasted in form of heat. The greater the degree of freedom in spintronic-based devices the more their usage in information processing. In contrast to a single value of charge of an electron, spin can assume values, +1/2 and −1/2 analogous to bits 1 and 0 in binary coding. For FM materials, application of the magnetic field can change the spin of electrons. Whereas, upon removal of the same magnetic field, the current spin values are retained. Thus, the information encoded remains stored without any requirement for power and no risk of demagnetization. Spintronic devices manipulate the electron spin in order to produce and control the charge currents besides interconverting the magnetic and electrical signals. In future spintronics-based devices have potential to overperform traditional semiconductor-based electronics owing to their ability to integrate several functionalities on a single platform namely, data storage and processing, sensing and logic operations. MRAM employs GMR or TMR and in order to write memory utilizing applied magnetic field that requires considerable power consumption. This can be resolved by utilizing ME materials. This is besides several other advantages that ME-based spintronic devices offer including speed of data processing, ultra-low power consumption and scalability. Thus, this role of ME materials has been remarkable in the spintronic device applications and has been discussed thoroughly in this chapter.

5.2 Spin-Dependent Conduction and Spin Manipulation

The spinorbit interaction results in spin current analogous to charge current flow. In spintronics, the information is encoded, stored and further transmitted using the spin of an electron. Figure 5.1 shows a FM metal band structure. Here, the energy band splitting is due to the spin i.e., minority spin and majority spin or spin down and spin up respectively. In 1936, Mott, for the first time proposed the role of spin in order to explain the resistivity of FM metals at the Curie temperature (s) [5]. Contrary to the non-magnetic metals, FM metals possess asymmetric band structure for spin-up and spin-down electrons. As a result, mobilities which depend upon electronic states are different for spin-up and spin-down electrons. Precisely at the Fermi energy level, the electrons exist in different spin states and show varying mobility and henceforth, different conduction response. Thus, with the application of identical electrical field, the spin-up and spin-down electrons behave differently. This implies both spin currents will possess different resistivities. In addition, application

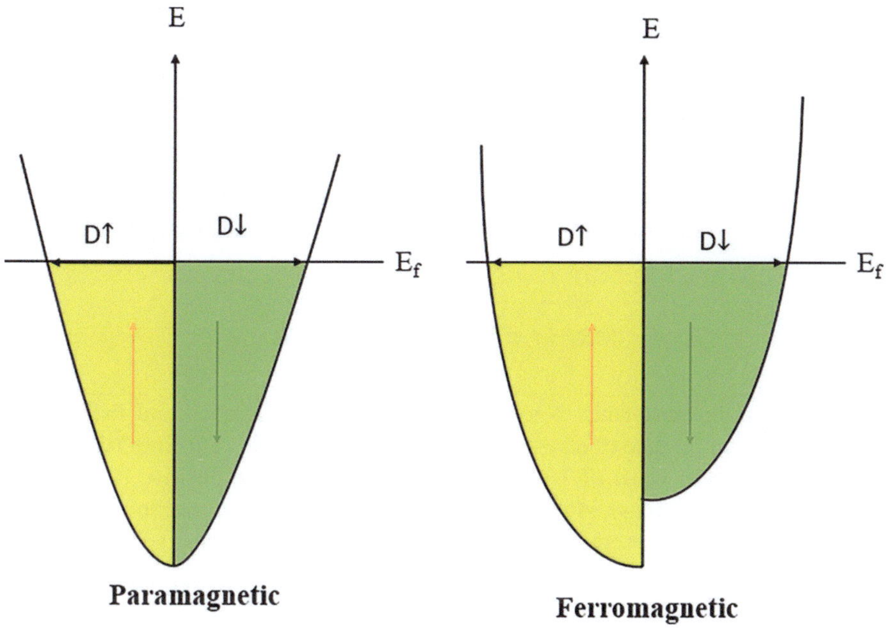

Fig. 5.1 Spin-up and spin-down electrons exhibit different behavior while placed in an identical electric field in the FM material. This is unlike the paramagnetic case whereby, both spin-up and spin-down electrons behave in a similar fashion

of magnetic field facilitates spin manipulation. This facilitates control of electric currents through the applied magnetic field.

In 1966, Albert Fert and Ian Campbell proposed a series of experiments to gain insight into spin-dependent conduction thereby providing a theoretical framework for the two-current model. One of the experiments involved study of spin-dependent scattering cross-section for Nickel doped with impurities. For e.g., Co-doped Ni exhibit a strong scattering for spin-down electrons, while Rh-doped Ni strongly scatter spin-up electrons. For Co- and Rh-co-doped Ni, both the spin-up and spin-down channels get strongly scattered, thereby resulting in an enhanced value of resistivity [6, 7]. However, for Co- and Au-doped Ni, strong scattering is observed for only one spin channel, while the other channel remains open. This gives rise to the concept of spin dependent resistivity or magnetoresistance. In the 'two current model' approach, electric current corresponding to spin-up and spin-down channels exist separately. The different current channels possess different resistivity which depends upon the spin scattering cross-section of a specific and identical impurity atom-doped magnetic element (Ni, etc.). In case of co-doping of impurities in magnetic Nickel, both the spin channels can get mixed together, whereby, both the spin-up and spin down electrons can get scattered, thereby reducing the scattering cross section. This results in increased resistivity. The net spin polarization, as a result of spin-mixing, also reduces in the FM material. Such a spin-mixing of channels is attributed to spin-orbit coupling whereby the electrons move at relativistic speeds at the Fermi level.

5.3 Magnetoresistance

Magnetoresistance is the ability of a single or a multiphase material (generally ferroelectrics) whereby its resistance exhibits a dependence on externally applied magnetic field. It can be classified in to categories viz., geometrical magnetoresistance, anisotropic magnetoresistance, giant magnetoresistance, colossal magnetoresistance, extraordinary magnetoresistance. Magneto-resistance (MR) can be quantified mathematically as:

$$\mathrm{MR}(\%) = \frac{R(0) - R(H)}{R(0)} \times 100 \tag{5.1}$$

where, $R(0)$ is the resistance when no magnetic field is applied and $R(H)$ is the resistance in the presence of externally applied magnetic field. This implies that MR can have maximum value of 100%. In addition, MR can be positive if resistance increases with an increase in magnitude of externally applied magnetic field and, negative if resistance decreases with an increase in magnitude of externally applied magnetic field.

5.3.1 Giant Magnetoresistance

Giant magnetoresistance (GMR) is observed in multiphase systems consisting of alternating interlayers of FM-nonmagnetic (conductive) materials. Here, 'giant' i.e., a large change in resistance, as compared to earlier observed results for metals, is observed with an externally applied magnetic field. Here, a spacer layer (few nm thick) is sandwiched between two magnetic layers as shown in Fig. 5.2. The first magnetic layer is characterized by a specific spin state whereby only one specific spin of electrons can pass through. Consequently, if the second magnetic layer is aligned with the first, then the corresponding spin channel can pass through the second layer and henceforth the crystal structure easily, thus, offering a very little resistance. However, if the second magnetic layer is misaligned with the first, then the spin channel gets hindered and electrons cannot easily pass through the second layer, consequently showing a very high electrical resistance. This can be understood using quantum mechanics. The electrons with different spin will also have different scattering rates. This is because the FM band structure splits consequently leading to different density of states for spin up and spin down electrons at the fermi level. Then according to the Fermi golden rule, the scattering rates for electrons are proportional to the 'density of states' of the state the electrons are scattered into i.e., fermi energy level. This implies that electrons with different spins will possess different scattering rates. Now, the two magnetic layers can be made parallel with application of large magnetic fields which can saturate the two layers with domains aligned parallel. Generally, the large and abrupt change in resistance is utilized to sense the magnetic field(s). GMR can be measured with two configurations namely, (i) Current-in-plane (CIP) GMR, and (ii) Current-perpendicular-to-plane (CPP) GMR geometry. The 'parallel current' moves parallel and in-between the two FM layers. This finds applications in magnetic read heads. In 'perpendicular current', one FM layer functions as the spin polarizer while the other as the detector. Here, for both geometries, MR is facilitated at the interface by spin-dependent scattering. In addition, spin accumulation at the interface also contributes to MR for CPP geometry. The strain-mediated ME coupling can also facilitate the control of magnetic anisotropy associated with FM layers. In this respect, we can also employ electric field to manipulate the magnetization direction of FM layers which results in an induced MR effect. One example is a spin valve structure composed of Ru/CoFe/Cu/CoFe/IrMn/Ru heterostructure deposited on a PMN-PT (011) substrate [8]. This finds applications in tunable-GMR-magnetic field sensor devices.

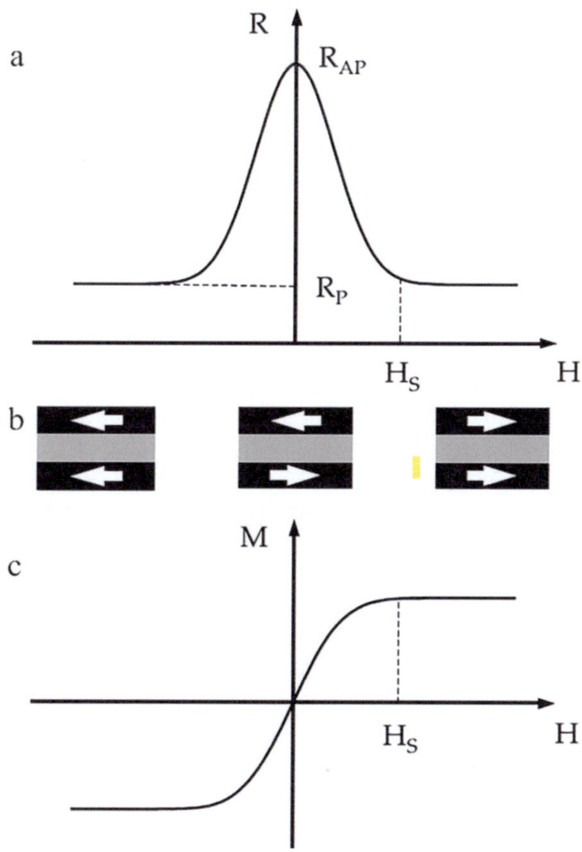

Fig. 5.2 Illustration of GMR effect, **a** shows resistance dependence on externally-applied magnetic field, **b** direction of magnetization of domains in the two magnetic layers separated by a non-magnetic spacer layer of a required specific thickness, and **c** magnetization response. Here, we observe that at the zero value of applied magnetic field, the magnetization direction of the two magnetic layers is aligned in antiparallel (AP) direction and we get a maximum value of resistance (R_{AP}), whereas with the applied magnetic field greater the saturated field (H_S), the magnetizations of the two magnetic layers are aligned in parallel direction and we observe a sharp reduction in the resistance values. Adapted with permission from Ref. [9]

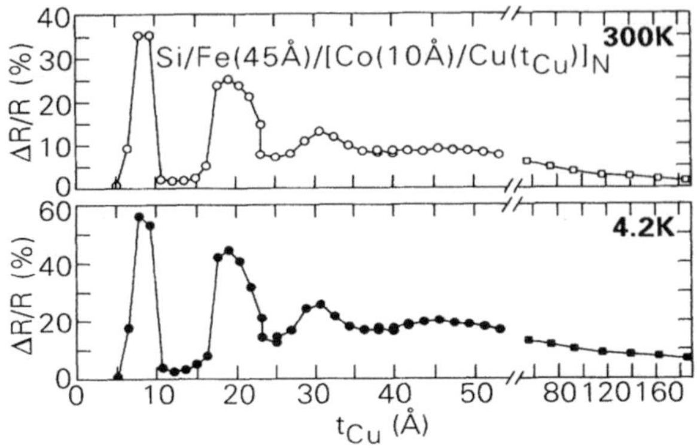

Fig. 5.3 Resistance dependence on the thickness of the spacer layer. Adapted with permission from Ref. [14]

5.3.2 *Antiferromagnetic Ordering*

Within a certain range of thickness of the spacer layer, the two adjacent magnetic layers will possess opposite magnetization. In addition, a sufficiently high magnetic field can force the magnetization of the two magnetic layers to point in the field direction. This facilitates the two resistance states in the multilayered heterostructures also known as superlattices. The superlattice structure comprise of a magnetic layer and an adjacent non-magnetic spacer layer arranged alternatively. For e.g. (001) Fe/(001) Cr [3], Co/Cu, Fe/Cu, Co/Cr, Fe/Cr, Co/Ru, etc. [10–13].

The GMR effect depends upon the thickness of the spacer layer as shown in Fig. 5.3. For instance, for Co/Cu multilayer heterostructures, the saturation magnetoresistance observed at room temperatures depends upon the Cu spacer layer thickness [14].

The maximum value of magnetoresistance is observed when the Cu spacer layer thickness is just 9 Å [14]. Such a high value of GMR correspond to the case when the neighboring 'Co' layers being coupled antiferromagnetically. Further, a very low value of resistance corresponds to the case when mean free path of electrons is large i.e., a large distance is traversed by electron between two consecutive scattering events. This means that the two neighboring 'Co' magnetic layers have their domains magnetized parallel to each other thereby making plausible an unhindered path for the electrons.

5.3.3 *Pseudo Spin Valves*

In some instances, we can employ different material compositions of the two magnetic layers, having their respective and different 'switching magnetic fields'. Then, as the external reverse magnetic field is applied, one of the two magnetic layers will switch before the other. This is as shown in Fig. 5.4. This results in the magnetic domains of the two magnetic layers to be aligned in the requisite antiparallel

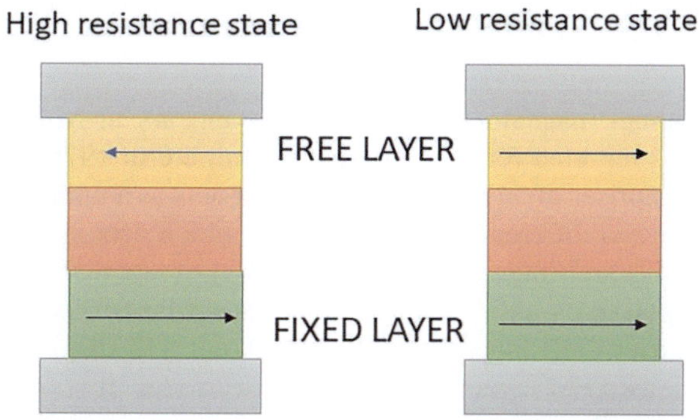

Fig. 5.4 Schematic depicting the structure of a pseudo spin valve whereby a spacer layer is sandwiched between two magnetic layers of different composition. One of the two magnetic layer is fixed while the other is free to move

direction. Such heterostructure compositions are referred to as pseudo spin valves. For e.g.: [Co/Cu/Fe/Cu] × N [15]. The high resistance state occurs when the two magnetic layers which are of different material but their domains are aligned parallel to each other under application of a magnetic field. This offers an unhindered path to electrons thereby resulting in a state of high resistance. Alternatively, the low resistance state corresponds to the case when the two magnetic layers comprising of different material composition have their magnetic domains aligned in an antiparallel direction under application of the magnetic field. This results in obstruction to the flow of electrons i.e., a small mean free path. This corresponds to a state of low resistance.

5.3.4 Spin-Valve Structures

Spin-valves have resulted the GMR devices to possess memory. Spin-valves are material layered structures consisting of a free layer and a fixed layer. The fixed layer (magnetically hard) is pinned and cannot move and henceforth is used as a reference for the measurement of magnetic moment. The fixed layer is made up of a pinning layer and a pinned layer. Usually, the free layer (magnetically soft) can be aligned even by a small magnitude of the applied magnetic field. Both the free and fixed layers are separated by a spacer layer. These structures are generally deposited on the substrates such as Si wafer. As depicted in Fig. 5.4, the free layer gets aligned in the direction of the applied magnetic field and acts as a soft layer. The fixed or pinned layer is an antiferromagnet and exhibits high magnetic coercivity consequently acts as a hard layer. Since both the layers have different coercivities, the soft layer will change its polarity at lower magnitude of the applied magnetic field as compared with the hard layer. Therefore, with appropriate magnitude of the applied magnetic field, the free/soft layer can alter its polarity thereby resulting in two distinct shows i.e., high and low resistance states. It is interesting to note that akin to the AFM in the spin-valve structures, the FM/FE interfaces of the ME materials in MFTJ also function similarly [16].

5.3.5 Tunnel Magnetoresistance

Tunnel magnetoresistance (TMR) appears in Magnetic Tunnel Junctions (MTJ) was reported for the first time by Julliere in 1975 Fe–Ge–Co junctions [17]. MTJ comprises of an insulating layer that acts as a tunnel barrier sandwiched in between the two different FM metal layers which acts as electrodes as shown in Fig. 5.5. MTJs are similar to spin valves (SVs) except that the metallic spacer layer in SV is replaced by an insulating tunnel barrier layer. The materials utilized for the tunnel barrier (TB) layer are AlO_x, MgO, etc. [18–20]. This TB is sandwiched between the two FM layers which acts as electrodes. It is observed that the tunneling resistance (R) of such junction devices is lesser when the magnetization in the two FM layers is parallel (Fig. 5.5a), whereas 'R' is comparatively more when the magnetizations of the two FM layers are in antiparallel direction (Fig. 5.5b). Then, the change in resistance with the change in the relative orientation of the two FM layers is termed as 'tunnelling magnetoresistance' (TMR) effect. The magnitude of TMR, as per Jullière's model, can be evaluated quantitatively by the magnetoresistance (MR) ratio i.e. [21],

$$\mathrm{MR} = \frac{R_{\mathrm{AP}} - R_P}{R_P} = \frac{G_P - G_{\mathrm{AP}}}{G_P} \tag{5.2}$$

where, R_P and R_{AP} are the resistance when magnetization of the two FM layers are parallel and antiparallel respectively. Also, G_P and G_{AP} are the conductance when magnetization of the two FM layers are parallel and antiparallel respectively. Also, in terms of polarization (P), TMR can also be expressed as:

$$\mathrm{TMR} = \frac{2P_1P_2}{1 - P_1P_2} \tag{5.3}$$

$$P_i = \frac{N_{Mi} - N_{mi}}{N_{Mi} + N_{mi}} \tag{5.4}$$

Here, N_{Mi} and N_{mi} are density of states corresponding to majority and minority spin states respectively. Also, we can describe conductance as $G_P \sim N_{M1}N_{M2} + N_{m1}N_{m2}$, $G_{AP} \sim N_{M1}N_{m2} + N_{m1}N_{M2}$. The TMR effect Fe–Ge–Co layers exhibits a MR of 14% at a temperature of 4.2 K [22]. Notably, the TMR effect was followed by the discovery of GMR effect in the late 1980s. In 1995, Moodera et al. explored GMR in CoFe/Al_2O_3/Co layers, which exhibited a MR of 18% at room temperature [23]. Thus, TB enclosed between two FM layers exhibit significant MR [24, 25]. Here, ME and MF materials have been effectively used as TB [26]. MFTJs comprise of a MTJ and tunneling barrier comprising ferroelectric films. This marks the co-existence of TMR and tunneling electroresistance (TER) effects. This also signifies a four-resistance state device whereby the resistance of such a device can be manipulated by application of both electric and magnetic fields. One example of a MFTJ device is a tunnel barrier comprising of a ferroelectric $BaTiO_3$ sandwiched between ferromagnetic $SrRuO_3$.

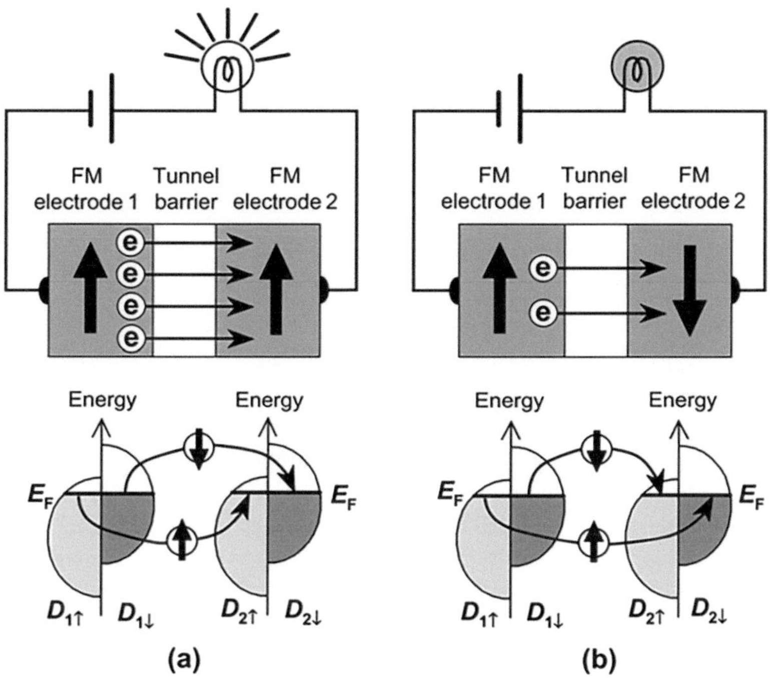

Fig. 5.5 **a** The tunneling resistance is lesser when the magnetization in the two FM layers is parallel, as compared with **b** where, the magnetization of the FM layers are in antiparallel direction. Adapted with permission from Ref. [22]

The ME coupling takes place at the FM/FE interface via direct i.e., spin exchange, strain, and charge coupling mechanism. However, it is also important to emphasize the role of 'magnetic' domain walls in the 'ferroelectric matrix' in facilitating ME coupling. The ME coupling at the interface influences the exchange splitting in spin bands of $SrRuO_3$ which in-turn changes the spin polarization and the TMR of the MFTJ device [27].

5.4 Spin Hall Effect

The first evidence of spin hall effect was found in 1974 for direct gap III-V semiconductors, InSb and Ge [28, 29]. The term, Spin Hall Effect (SHE) was, however, introduced by Hirsch in 1999. It is a transport phenomenon which was first theoretically predicted in 1971 by Dyakonov and Perel. It finds origin in charge-spin coupling due to spin-orbit interaction. Here, an electric current carrying conductive material exhibits spin accumulation at its lateral boundaries. Also, the spin directions are opposite at the opposite boundaries as shown in Fig. 5.6. In Hall effect, magnetic field is applied perpendicular in direction to the electric current passing

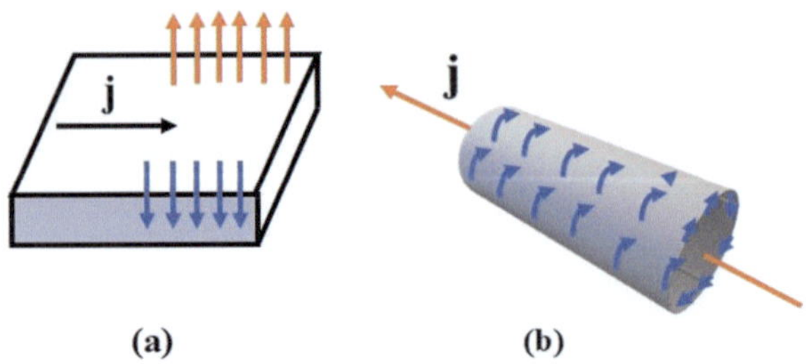

Fig. 5.6 Illustration of the spin hall effect. **a** An electric current effectuates spin accumulation at the lateral boundaries of the material. **b** An electric current flowing through a cylindrical conductor produces a spin effect whereby spins are wind around the conductor surface. Redraw in the impression of Ref. [30]

through a conductor. This results in production of a Hall voltage or a potential difference in transverse direction to both electric current and applied magnetic field. The Hall voltage is developed as there is accumulation of charges of opposite sign at the conductor boundaries as a result of Lorent's force due to the applied magnetic field. The SHE is also similar to the Hall effect except, firstly, the spin accumulation in SHE does not need application of the magnetic field. In fact, the magnetic field perpendicular to spin accumulation can destroy spin polarization and SHE. Secondly, spin polarization is subject to spin relaxation and is facilitated by the spin layers whose width is determined by the spin diffusion length which is typically ~1 μm.

Interestingly, bulk-SHE is known to induce bilinear magnetoelectric resistance (BMER). In BMER, the non-linear MR is known to scale in a linear fashion with applied electric and magnetic field simultaneously. BMER has been typically observed on surface of 3D topological insulators, Rashba 2D electron gases, and heterostructure based on AFM materials [31].

5.5 Spin Current

Spin accumulation results in corresponding spin current which is represented by a tensor quantity i.e., q_{ij}, where, the first index represents the flow direction, and the second index represents the flowing component of spin. If 'n' is the concentration of electrons i.e., number of electrons per unit volume, acquire a velocity v in $+x$ direction and are spin polarised along z direction. Then, non-zero component of spin polarization current density is a tensor q_{ij}, $q_{xz} = nv$. For spin ½ particles, $q_{xz} = \frac{1}{2}nv$, however, we usually omit '½' factor in order to avoid numerous '½' and '2' factors in other places. The charge density and spin current flow density can be represented as:

$$q_i^{(0)} = -\mu n E_i - D\frac{\partial n}{\partial x_i} \quad (5.5)$$

$$q_{ij}^{(0)} = -\mu n E_i P_j - D\frac{\partial P_j}{\partial x_i} \tag{5.6}$$

Here, μ is mobility, D is the diffusion constant, n is number of electrons per unit volume, E_i is ith component of electric field vector **E**, P_j is the jth component of spin polarization density vector **P**. Here, Eq. (5.1) represents electron's drift-diffusion relation, and (5.2) represents spin current density of polarized electrons. Additionally, since the spin can be transported through electron flow, the spin-orbit interactions facilitate but are not necessary for spin current flow. Further, considering spin-orbit interaction, the above two expression for currents after corrections can be re-written as for an isotropic material and considering inversion symmetry in tensor form as follows:

$$q_i = q_i^{(0)} + \gamma \varepsilon_{ijk} q_{jk}^{(0)} \tag{5.7}$$

$$q_{ij} = q_{ij}^{(0)} + \gamma \varepsilon_{ijk} q_k^{(0)} \tag{5.8}$$

where, we use Einstein's summation convention for the tensors, q_i and q_{ij} are the final corrected values for current expression. Also, γ is dimensionless and is directly proportional to the strength of spin-orbit interaction, ε_{ijk} is an antisymmetric tensor i.e., $\varepsilon_{xyz} = -\varepsilon_{yxz} = \varepsilon_{yzx} = -\varepsilon_{zyx} = \varepsilon_{zxy} = -\varepsilon_{xzy} = 1$, and the other values being '0'. Thus, the tensor q_{xy} symbolizes the spin current which facilitates the flow of charge carriers in z direction. In other words, the charge carriers flowing in z direction induce spin current(s) q_{xy} (q_{yx}).

5.6 Spintronic Devices

They utilize manipulation of spin polarized electrons to get excess of either spin up or/and spin-down electrons. This non-equilibrium population of spin up and spin down electrons exist within a characteristic spin lifetime τ and over a certain distance in which the non-equilibrium spin polarized electrons can travel i.e., spin diffusion length (λ). Spintronic devices can be further classified based on the material system chosen namely, (i) all metallic spin transistors which also includes spin valves, (ii) MTJ and MFTJ-based devices, (iii) FM-semiconductor heterostructure spintronic devices, and (iv) organic material-based spintronic devices.

5.6.1 Spin-Based Logic Devices

The spin-based non-volatile digital logic devices have attracted immense potential owing to its low power consumption technologies. As shown in the Fig. 5.7a, X and

Y are two logic gate inputs to NAND gate inputs. Also, A = B = 1 whose values can, however, be set for different logical operations by adjusting their magnetic alignments. The output of this NAND gate is represented by the transient current $I_M(t)$. The spin accumulation of the electrons is directly proportional to the amplitude of $I_M(t)$. Amid magnetization rotation, transient current passes through M (probe contact) resulting in perturbing of the potential level in M. This causes the capacitor to reinforce the steady state i.e., $I_M = 0$. This transient current requires to be converted into binary logic '0' and '1' voltage signal. Here, V_{dd} is the input supply voltage. Further, as shown in Fig. 5.7b, All-spin logic (ASL) switch has immense potential for low power consumption electronic systems. The low-power consumption for ASL-based systems is attributed to the non-local spin currents as shown in the Fig. 5.7b. Here, without the requirement of spin-to-charge conversion, the communication signals can be selectively routed through appropriate channels in between the specified input and output magnets [32].

We require spintronic logic device which is scalable, high logic density and has low-power consumption employing spin-orbit transduction along with ME switching [35]. The magnetoelectric spin-orbit (MESO) involves two features namely, FE/ME

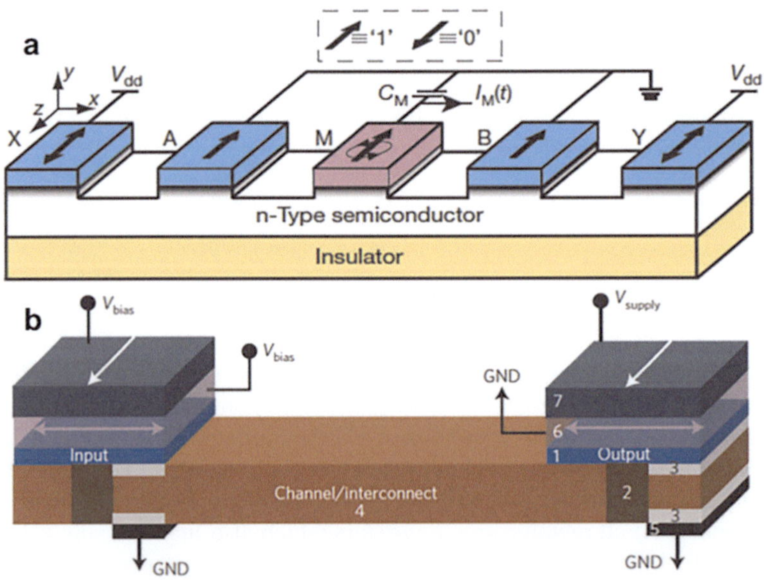

Fig. 5.7 **a** An illustration representing spin logic device depicting NAND gate. **b** represents an "all-spin logic" (ASL) switch whereby, information is facilitated in between the two nanomagnets utilizing the non-local origin spin currents. Here, we employ voltages i.e., V_{supply} to facilitate application of a large torque and V_{bias} to exert a small magnitude torque in order to receive and send information signals. The two spin torque(s) decide the output. Here, '1' is the magnetic free layer; '2' and '3' are isolation and tunneling layers respectively; '4' is the spin coherent channel; '5' is a contact; '6' and '7' are spacer and magnetic fixed layers respectively. Adapted with permission from Ref. [33, 34]

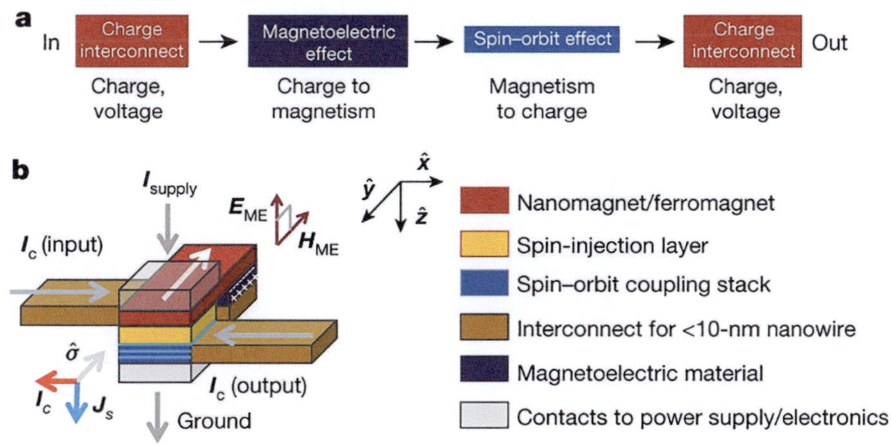

Fig. 5.8 **a** Flowchart showing a charge-input and charge-output logic device and **b** a MESO device fabricated using ME capacitor with a topological material and comprises the spin-injection layer which injects spin from FM to the ME material layer, interconnecting made up of a conductive material and finally contacts to ground and power supply. Adapted with permission from Ref. [35]

switching besides spin-to-charge topological conversion. As shown in Fig. 5.8, the specified MESO device can invert the input charge i.e., positive voltage and current and the output of the logic device is obtained as inverted.

5.6.2 Spintronic Transistors

Spintronic transistors are composed of two-terminal MR devices as their basic building blocks. The current perpendicular to plane geometry of the spin valve is utilized to synthesize the spin transistors. Here, the FM layers comprise the FM transition metals and Heusler alloys whereas the ultrathin spacer layer consists of non-magnetic metals and non-magnetic transition metal elements.

Notably, in 1989 Datta and Das proposed an electron spin-based device analogous to an electro-optic cell [36]. This then later formed basis for spin Field Effect Transistor. In a wave modulator, as shown in Fig. 5.9, the electro-optic cell is placed between the polarizer and analyzer. It, then acts as a light transmitter or light blocker whereby, the electro-optic unit rotates the plane of polarization of the incoming polarized light from the polarizer, provided there is an application of requisite bias voltage.

Datta and Das then proposed an analogous device utilizing property of electron spin rather than light polarization as shown in Fig. 5.10. FM material is the source of majority spin whereby, the injected electrons possess a majority spin (up or down) facilitated by the existence of non-equilibrium between the two spin states. This is

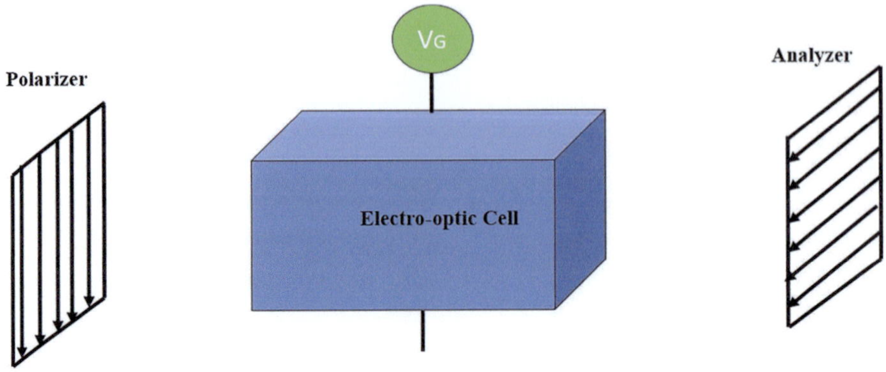

Fig. 5.9 Schematic of a wave modulator: electro-optic cell supplied with a proper bias voltage is placed between the polarizer and analyzer. Magnitude of voltage bias manipulates the wave modulator working as a wave transmitter or wave blocker

because the FM source contact possesses majority spins up or down which determines their magnetization direction. The majority spin (spin up or spin down) current then flows through the semiconductor channel. The source contact is analogous to a polarizer in a wave modulator device as the former injects a majority spins into the semiconducting channel. These majority spin electrons pass through 2DEG i.e., two-dimensional electron gas inside the semiconducting channel. Also, the direction of the majority spins via the Gate bias voltage can be aligned in the same or opposite direction towards the 'FM' drain contact corresponding to the high and low conductance of the device respectively. The magnitude of gate voltage determines the spin alignment. The phenomenon where the electron's spin precession angle is dependent upon the strength of applied voltage is known as 'Rashba' effect.

When the FM drain contact's magnetization direction is aligned parallel to the orientation of the majority spin electrons at the drain channel, the spin current flows through the drain and the spin Field Effect Transistor is in 'ON' state. Similarly,

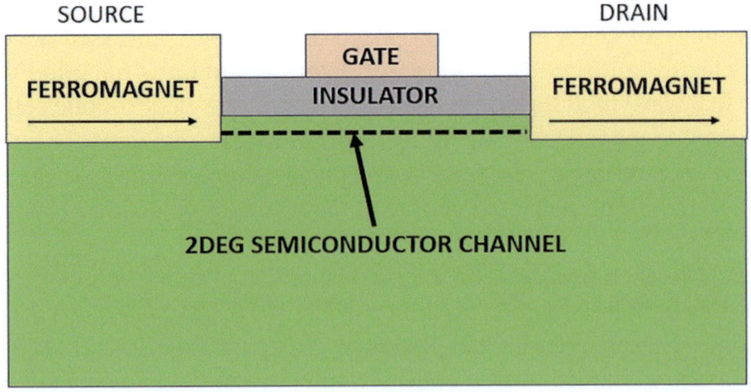

Fig. 5.10 Schematic of a spin-based field effect transistor

the FM drain contact's magnetization direction is aligned opposite or anti-parallel to the orientation of the majority spin electrons at the drain channel, the electron spin current does not flow through the drain and the spin Field Effect Transistor is in 'OFF' condition. This preferred spin alignment in the semiconducting channel can be induced by changing the Gate Voltage that effectuates angle of spin precession of electron via 'Rashba' effect. This implies that changes in gate voltage correspond to the changes in transistor 'ON' and 'OFF' states.

The requirements for a spin field effect transistor are, (i) the electron spin-current being injected from FM material source into the semiconducting channel, (ii) electron spin-current being further transported through the semiconducting channel, and (iii) finally, the electron spin current being detected and further transported in to the FM drain contact. For the first requirement, it is important to understand that the FM source material consists of spin oriented in a certain direction that forms the magnetization direction. These spin-polarized electrons cannot be injected into the semiconducting channel by application bias voltage to the FM-semiconductor channel contact. This happens because of the conductivity mismatch. The applied voltage appears across the semiconductor itself due to the higher conduction of the FM metal. This is however, overcome by inserting a thin layer of an insulating material in between the FM metal and semiconductor. Consequently, this provides a tunnel resistance and allows the injection of spin-dependent charge carriers into the semiconducting channel. This transport of spin-polarized current is as shown in Fig. 5.11 by arrow '1'. Further, the second requirement necessitates the transport of charge carriers through the semiconducting channel as shown by arrow '2' in Fig. 5.11. It is important that the spin direction is preserved during the entire transportation distance across the semiconducting channel. However, diffusion of spin-polarized electrons during the transportation process results in scattering effectuating the spin-polarized charge carriers to lose their spin direction after a brief distance. Thus, it is pertinent for electrons to suffer the minimum scattering in the semiconductor channel i.e., while traversing the path between the source and the drain contact. Thirdly, we require the spin-polarized charge carriers to be further transported to the FM drain contact. As explained in the first requirement, there exists a conductivity mismatch between the semiconductor channel and FM drain contact. These further necessities a thin insulating layer to be placed in between the two contacts. The corresponding spin-polarised current being injected into the drain is shown by arrow '3' in Fig. 5.11 Here, arrow '4' represents the electrons which have lost their property spin direction as a result of backscattering.

With this basic understanding of spin field effect transistors(s-FET), numerous possible device structures have been explored by the researchers worldwide. For instance, a Silicon/semiconductor on Insulator (SOI) s-FET has been studied [37, 38]. These thin structures provide an improved device performance as well.

With this basic understanding, we now move forward towards ME spin field-effect transistors (ME spin FET) as shown in Fig. 5.12. Here, the voltage controlled ME composite generate a magnetic field which effectuates the spin-orbit splitting of the electronic energy bands of the thin semiconductor layer. This results in a transverse and also an overall spin hall current. Thus, a strong spin-orbit coupling present in

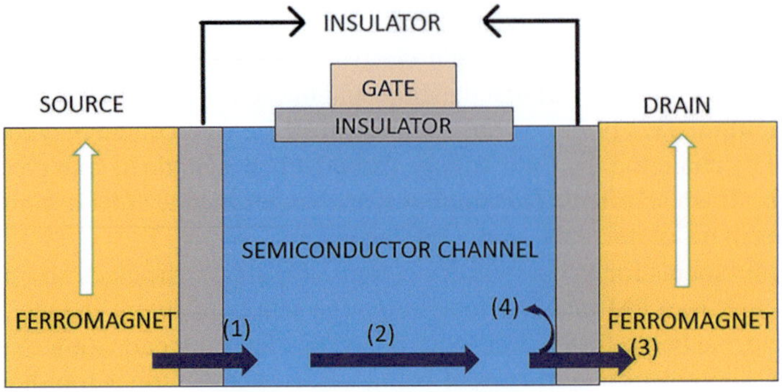

Fig. 5.11 Illustration of a spin-dependent field effect transistor

the semiconductor channel will facilitate a corresponding spin hall current. Also, Fig. 5.12 shows two FM source channels which are associated with characteristic drain-source voltage besides current difference. In order to configure this device as a logic gate, we provide two inputs namely voltage at the gates and magnetization direction in the semiconducting channel which is a result of boundary polarization. Then we describe the up and down direction of magnetization as logic '0' and '1' respectively. In addition the two polarization directions which are switched by voltages 0 and 0.1 V are designated as logic '0' and '1'. The obtained logic states are non-volatile i.e., the state is retained in the memory of the logic even when the power of the device is switched off [39].

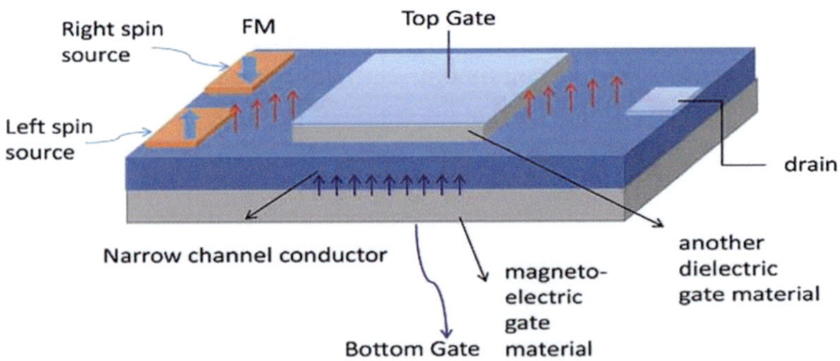

Fig. 5.12 Illustration of a ME spin-FET multiplexer (spin-MUX) with FM source as contacts. Bottom gate comprises of the ME composite upon which lies a thin conductor or semiconductor that can be polarized up or down. Adapted with permission from Ref. [39]

5.7 Conclusions

Spintronic devices utilize spin of an electron which can be manipulated by the application of magnetic field. Unlike non-magnetic metals, FM metals possess asymmetric band structure for spin-up and spin-down electrons. This property is utilized to fabricate spintronic devices. GMR involves large and abrupt change in resistance upon application of magnetic field and thus, has been often utilized in magnetic sensor devices. Then, there are spin valve structures which are materials having layered structures of a free layer and a fixed layer separated by a spacer layer. The fixed layer is magnetically hard and is pinned and cannot move and henceforth is used as a reference for the measurement of magnetic moment. The free layer, on the other hand is magnetically soft and can be aligned by a small magnitude of magnetic field. Also, SHE finds applications in spin current detectors and generators, spin transistors, optical monitoring, precision metrology, nanophotonic devices, etc. Further, ME-based spintronic devices have been studied in detailed for spin-based non-volatile digital logic devices, spin Field Effect Transistors, etc. which find applications in data storage, medical industry (cancer detection, neuromorphic computing, etc.) owing to their low power consumption in electronics industry.

References

1. Wolf SA, Awschalom DD, Buhrman RA, Daughton JM, von Molnár S, Roukes ML, Chtchelkanova AY, Treger DM (2001) Spintronics: a spin-based electronics vision for the future. Science 294:1488–1495
2. Fert A (2008) Nobel lecture: origin, development, and future of spintronics. Rev Mod Phys 80:1517–1530
3. Baibich MN, Broto JM, Fert A, Nguyen Van Dau F, Petroff F (1988) Giant magnetoresistance of (001)Fel(001)Cr magnetic superlattices. Phys Rev Lett 61:2472–2475
4. Bruno P, Chappert C (1991) Oscillatory coupling between ferromagnetic layers separated by a nonmagnetic metal spacer. Phys Rev Lett 67:1602
5. Mott F (1936) Proc R Soc London Ser A 153:699
6. Hirsch JE (1999) Spin hall effect. Phys Rev Lett 83:1834–1837
7. Fert A (2008) The present and the future of spintronics. Thin Solid Films 517:2–5
8. Wang L, Hu Z, Zhu Y, Xian D, Cai J, Guan M, Wang C, Duan J, Wu J, Wang Z, Zhou Z, Jiang Z-D, Zeng Z, Liu M (2020) Electric field-tunable giant magnetoresistance (GMR) sensor with enhanced linear range. ACS Appl Mater Interfaces 12(7):8855–8861. https://doi.org/10.1021/acsami.9b20038
9. Tsymbal EY, Pettifor DG (2001) Perspectives of giant magnetoresistance, solid state physics, vol 56. Academic Press, pp 113–237
10. Parkin SSP, More N, Roche KP (1990) Oscillations in exchange coupling and magnetoresistance in metallic superlattice structures: Co/Ru, Co/Cr, and Fe/Cr. Phys Rev Lett 64:2304
11. Parkin SSP, Li ZG, Smith DJ (1991) Giant magnetoresistance in antiferromagnetic Co/Cu multilayers. Appl Phys Lett 58:2710–2712
12. Grünberg PA (2008) Nobel lecture: from spin waves to giant magnetoresistance and beyond. Rev Mod Phys 80:1531–1540

13. Hecker M, Thomas J, Tietjen D, Baunack S, Schneider CM, Qiu A, Cramer N, Camley RE, Celinski Z (2003) Thermally induced modification of GMR in Co/Cu multilayers: correlation among structural, transport, and magnetic properties. J Phys D: Appl Phys 36:564–572
14. Parkin SSP, Bhadra R., and Roche K.P. (1991) Oscillatory Magnetic Exchange Coupling through Thin Copper Layers. Phys. Rev. Lett. 166:2152–2155
15. Dieny B, Prejbeanu IL, Garello K, Gambardella P, Freitas P, Lehndorff R, Raberg W, Ebels U, Demokritov SO, Akerman J, Deac A (2020) Opportunities and challenges for spintronics in the microelectronics industry. Nat Electron 3:446–459
16. Quindeau A, Fina I, Marti X et al (2015) Four-state ferroelectric spin-valve. Sci Rep 5:9749. https://doi.org/10.1038/srep09749
17. Julliere M (1975) Tunneling between ferromagnetic films. Phys Lett 54A:225
18. Miyazaki T, Tezuka N (1995) BGiant magnetic tunneling effect in Fe = Al_2O_3 = Fe junction. J Magn Magn Mater 139(3):L231–L234
19. Wang D, Nordman C, Daughton JM, Qian Z, Fink J (2004) B70% TMR at room temperature for SDT sandwich junctions with CoFeB as free and reference layers. IEEE Trans Magn 40(4):2269–2271
20. Butler WH, Zhang X-G, Schulthes TC, MacLaren JM (2001) BSpin-dependent tunneling conductance of FejMgOjFe sandwiches. Phys Rev B 63(5):054416/1–054416/12
21. Žutić I, Fabian J, Sarma SD (2004) Spintronics: fundamentals and applications. Rev Mod Phys 76:323
22. Yuasa S, Djayaprawira DD (2007) Giant tunnel magnetoresistance in magnetic tunnel junctions with a crystalline MgO(0 0 1) barrier. J Phys D: Appl Phys 40:R337–R354
23. Moodera JS, Kinder LR, Wong TM, Meservey R (1995) Large magnetoresistance at room temperature in ferromagnetic thin film tunnel junctions. Phys Rev Lett 74:3273
24. Parkin SSP, Kaiser C, Panchula A, Rice PM, Hughes B, Samant M, Yang S-H (2004) BGiant tunnelling magnetoresistance at room temperature with MgO(100) tunnel barriers. Nature Mater 3:862–867
25. Lee YM, Miura K, Hasegawa H, Tsunoda M, Matsukura F, Ohno H, B (2008) Tunnel magnetoresistance of 604% at 300 K by suppression of Ta diffusion in CoFeB/MgO/CoFeB pseudo-spin-valves annealed at high temperature. Appl Phys Lett 93:082508/1–082508/3
26. Béa H et al (2006) Appl Phys Lett 88:062502
27. Niranjan MK, Burton JD, Velev JP, Jaswal SS, Tsymbal EY (2009) Magnetoelectric effect at the $SrRuO_3$–$BaTiO_3$ (001) interface: an ab initio study. Appl Phys Lett 95:052501. https://doi.org/10.1063/1.3193679
28. Chazalviel JN, Solomon I (1972) Experimental evidence of the anomalous hall effect in a nonmagnetic semiconductor. Phys Rev Lett 29:1676
29. Chazalviel J-N (1975) Spin-dependent hall effect in semiconductors. Phys Rev B 11:3918
30. Dyakonov MI, Khaetskii AV (2008) Spin physics in semiconductors, chapter spin hall effect, Springer series in solid state sciences, vol 157. Springer, Berlin, Heidelberg, pp 211–243. ISSN 0171-1873. https://doi.org/10.1007/978-3-540-78820-1_8
31. Kim DJ, Kim KW, Lee K et al (2024) Spin Hall-induced bilinear magnetoelectric resistance. Nat Mater 23:1509–1514. https://doi.org/10.1038/s41563-024-02000-0
32. Ahn E. C., 2D materials for spintronic devices, npj 2D Materials and Applications, 4, 17, 2020, https://doi.org/10.1038/s41699-020-0152-0
33. Dery H, Dalal P, Sham L (2007) Spin-based logic in semiconductors for reconfigurable large-scale circuits. Nature 447:573
34. Behin-Aein B, Datta D, Salahuddin S, Datta S (2010) Proposal for an all-spin logic device with built-in memory. Nat Nanotechnol 5:266
35. Manipatruni S, Nikonov DE, Lin CC et al (2019) Scalable energy-efficient magnetoelectric spin–orbit logic. Nature 565:35–42. https://doi.org/10.1038/s41586-018-0770-2
36. Datta S, Das B (1990) Electronic analog of the electro-optic modulator. Appl Phys Lett 56:665–667
37. Sugahara S (2006) Perspective on field-effect spin transistors. Phys Stat Sol 12:4405–4413
38. Datta S (2018) How we proposed the spin transistor. Nat Electron 1:604

39. Dowben PA et al (2018) Towards a strong spin-orbit coupling magnetoelectric transistor. IEEE J Exploratory Solid-State Comput Dev Circuits 4:1–9. https://doi.org/10.1109/JXCDC.2018.2809640

Chapter 6
Advances in Magnetoelectric Devices and Applications in Surface Acoustic Wave Device Technology

6.1 Introduction to ME Sensor Devices

The ME properties in a sensor device are governed by the elastic interaction which occurs between the magnetostrictive and piezoelectric phases in the composites. Further, depending upon the configuration of the magnetostrictive and piezoelectric layers, the operational modes of the sensor device can be classified as L-L, L-T, T-T and C–C where "L", "T", and "C" refers to the longitudinal, transverse, and circumferential modes respectively. The value of ME voltage coefficient (α_{ME}) is determined by the materials used, interface region, operational mode, and the dc magnetic bias field. Laminated structures exhibit α_{ME} up to ~20 V/cm Oe for the frequencies ranging from 1 Hz to 1 kHz. It is important to mention that α_{ME} exhibits extra-ordinary high value ($\times 100$) at the electromechanical resonance (EMR) frequency. Reportedly, α_{ME} of more than 1100 V/cm Oe at EMR for Metglas/$Pb(Mg_{1/3}Nb_{2/3})O_3$–$PbTiO_3$ (PMN-PT) was observed [1]. Notably, passive operation of a ME device corresponds to its operation at frequencies less than the EMR frequency. This facilitates a low power consumption and is a suitable for long-term usage of a battery-operated ME sensor. Usually, a low power consumption of 10–20 μW is observed for ME laminate structures [2]. The ME laminate composites have been utilized to fabricate ME-multifunctional devices owing to difficulty in obtaining interconnectivity between the different phases of the composite structure. Laminate composites fabricated as layered structures comprising of magnetostrictive/piezoelectric are known to exhibit high ac magnetic field sensitivity [3]. The ME laminate composites comprise of piezoelectric $Pb(Zr,Ti)O_3$, magnetic ferrites like $CoFe_2O_4$, $NiFe_2O_4$ etc., magnetic alloys such as Terfenol-D, Ni, Metglas, and piezoelectric composites like $Pb(Mg_{1/3}Nb_{2/3})O_3$–$PbTiO_3$ etc. The ME devices can be of very high sensitivity i.e. extremely low noise 10^{-12} $THz^{-1/2}$ at a frequency of 1 Hz. The control or manipulation of electric polarization by application of the magnetic field is known as direct ME coupling. The direct ME coupling finds applications in numerous multifunctional devices such as ME-based magnetic field and current sensors, gyrators, energy

M. Kumar and S. Dabas, *Multiferroic and Magnetoelectric Materials*, Materials Horizons: From Nature to Nanomaterials, https://doi.org/10.1007/978-981-95-4391-5_6

harvesting devices, magnetic nanoparticles detection etc. [4]. Magnetic field sensing also possess immense application in the biomedical field. For instance, the magnetic field signals generated by the brain can be detected successfully by a ME magnetic field sensing device. Similarly, ME gyrator can be used to convert the various network elements into their corresponding duals and vice versa, such as inductor into capacitor, resistor into admittance, voltage into current source and vice-versa. Further, the manipulation of magnetization by the application of electric field is known as converse ME coupling. It encompasses manipulation of spin waves by application of electric field as well as well as manipulation of permeability by the application of the electric field. Examples of converse ME coupling include spintronics, MERAM devices, voltage adjustable inductivities and filters, bandpass filters, phase shifters, mechanical antennas, resonators etc. Memory devices such as MERAM can utilised to store binary data which correspond to "stable magnetic states" and also to read data by detecting the resistance change of the memory element. Electric field tuned ME-VTIs have been utilized in numerous circuit designs. Phase shifters are generally utilized to change the phase of the electromagnetic signal. The ME-based Surface Acoustic Waves (SAW) devices are based on a transducer and find applications in delay lines, magnetic skyrmions, magnetic resonance, etc. SAW-based devices are generally known to be utilized in the electromechanical systems which are inherent part of sensing devices and in the signal processing. Here, all these applications are discussed in-depth in this chapter.

6.2 Applications of Direct ME Coupling

6.2.1 Magnetic Field ME Sensor Device

A magnetic field sensor can detect any change in the magnetic field from the surrounding environment [5]. They find applications in measurement of the Earth's magnetic field, meteorology studies, detection of weapons, in the security systems, etc. The externally applied ac magnetic field interacts with the ac magnetic field (to be detected). This interaction results in a linear pseudo-piezomagnetic behavior which can modulate the electrical signals of the magnetic sensor device. The linear ME voltage coupling coefficient is defined as:

$$\alpha_H^V = \frac{\partial E}{\partial H} = \frac{1}{t}\left(\frac{\partial V}{\partial H}\right) \tag{6.1}$$

where, E is the induced electric field, H is the applied magnetic field (which is to be detected) and 't' is the thickness of the material for single phase multiferroics or thickness of the FE layer component in case of laminate structure materials. Integration this relation results in $V = \alpha_H^V Ht$. Here, the output voltage is directly proportional to the strength of the ME coupling coefficient. Thus, single phase multiferroics exhibit

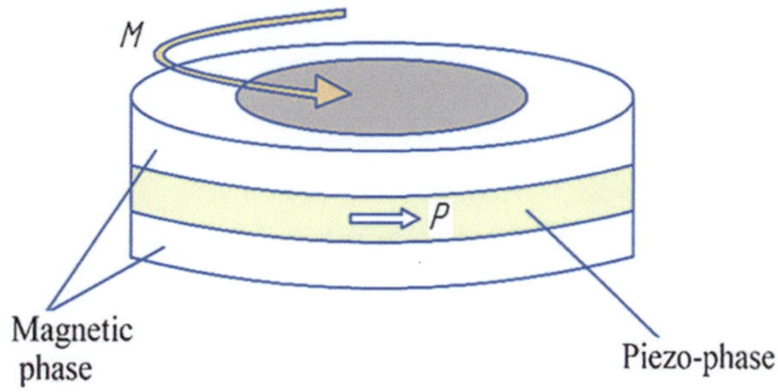

Fig. 6.1 A ME magnetic field sensor device comprising C–C layered structure comprising of Terfenol-D/PXT/Terfenol D layers. Adapted with permission from Ref. [2]

very low output voltage at extremely low temperatures on account of their lower ME coupling coefficient. Hence, the sensing device is usually composed of composite ME materials or laminate structures. These materials exhibit strain-mediated ME coupling coefficient.

They comprise of a circularly polarized piezoelectric phase of PZT i.e., $Pb(Zr,Ti)O_3$ sandwiched between the two circularly magnetized magnetostrictive phases of Terfenol-D i.e., ($Tb_{1-x}Dy_xFe_{2-y}$) as shown in Fig. 6.1 [3]. Such a structure results in enhanced sensitivity of 10^{-9} T. However, we get a large displacement field of 500 Oe.

6.2.1.1 Bulk Laminate

The bulk ME sensor device can detect magnetic fields of more than 1 to 10 pT at a frequency of 1 Hz. They are composed of a piezoelectric phase (($Pb(Zr,Ti)O_3$, $Pb(Mg,Nb)O_3$–$PbTiO_3$ etc.) and piezomagnetic phase (FeCoSiB, FeBSiC Terfenol-D, Ni-Ga, Fe-Ga etc.). The magnetic and FE phases in the composite structures can be composed in various ways and expressed through the connectivity schemes for a two-phase composite system as shown in Fig. 6.2. For instance, the "0-3" scheme implies a particulate composite consisting of single-phase nanoparticles embedded into the matrix of another phase. Similarly, in the 1-3 connectivity scheme we have vertical heterostructures of fibres/rods embedded in a matrix of another phase. The 2-2 connectivity scheme represents laminated horizontal "film" structures. Selection of a connectivity scheme depends upon the specific functional properties including electrical, magnetic, or mechanical properties required for a designated output. The 0-3 connectivity scheme is associated with leakage current problems which can be avoided by choosing a 2-2 connectivity scheme i.e., laminate structure.

Dong et al. have fabricated 2-2 type magnetostrictive Terfenol D-piezoelectric $Pb(Mg_{1/3}Nb_{2/3})O_3$–$PbTiO_3$ laminate structure [6]. The piezoelectric $0.7Pb(Mg_{1/3}Nb_{2/3})O_3$–$0.3PbTiO_3$ is sandwiched between two $Tb_{1-x}Dy_xFe_{2-y}$ i.e., terfenol-D layers. This is a L-T laminate whereby the piezoelectric and piezomagnetic layers are coupled together with stress (S(z)) and strain (T(z)). This is because the applied magnetic field along the length of the magnetostrictive layer effectuates

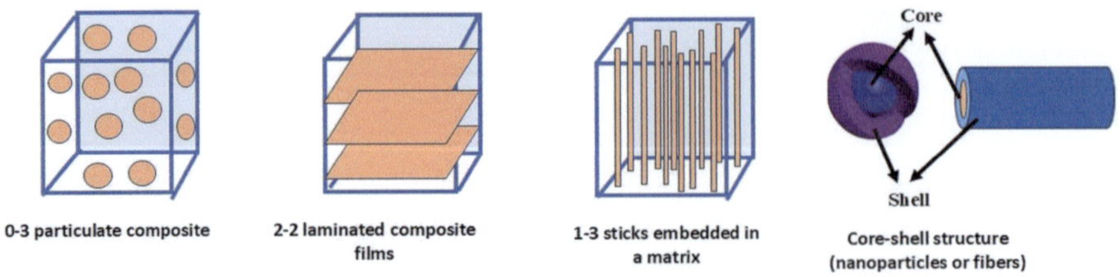

Fig. 6.2 Connectivity schemes in ME composites

the piezoelectric layer into forced oscillations in the very same direction. Here, a low value of ME coupling coefficient of 2.2 V/cm Oe was observed. Later studies reported a ME coupling coefficient of 22 V/cm Oe which is suitable for magnetic sensor devices [6–8].

6.2.1.2 Thin Film Laminate

Li et al. developed a Nanoelectromechanical system (NEMS) whereby a magnetic field sensor comprising of a piezoelectric layer (AiN) is sandwiched between a magnetostrictive layer (FeGaB) and interdigital electrodes (silicon strips) [9]. Here, FeGaB film of thickness 250 nm is deposited through sputtering on silicon strips having dimensions 35 mm × 3.5 mm × 150 μm. Finally, the response of FeGaB/AiN magnetic field sensing device was reported at a resonant frequency of 225.8 MHz with a DC bias field of 12 Oe. With a suitable cantilever arrangement, the resonant frequency can be managed to be in a specific range of 100 Hz to 100 kHz. The advantage of such thin film cantilever arrangement as compared with bulk composite-based magnetic field sensors is the considerable reduction in the size of the sensor. In addition, we can obtain a unidirectional magnetic field sensing which facilitates the vector field measurements. The Limit of Detection (LOD) of the magnetic field is determined by the noise signal generation in the absence of the magnetic field (to be detected). Reduction in the sources of external vibration increases sensitivity of the magnetic sensor devices [10].

6.2.1.3 Bio-medical Applications of Magnetic Field Sensor

In the biomedical field, we come across Magnetoencephalography (MEG), magnetocardiography (MCG) and magnetomyography (MMG) which is a technique to sense magnetic signals produced by brain [11]. Özden et al. fabricated thin film cantilever bio-magnetic sensors comprising of a substrate layer of silicon which is 300 μm thick, 26.25 mm in length, and 2.45 mm in width, piezoelectric aluminum nitride (AlN) layer which is 20 μm thick and piezomagnetic FeCoSiB layer which is 20 μm thick [12]. Three such fabricated sensors are placed mutually perpendicular in order to obtain a vector field sensing device. This is then applied to the 'head model' comprising of three concentric spherical sections representing the inner white matter, middle represents the skull, and outer most layer represents the skin. The electric dipoles present inside the "head structure" are responsible for generating EM radiations. This EM field reaches outside the "head structure" and finally reaches the fabricated sensor device where it is detected. The LOD for a biomedical sensor device corresponds to the smallest value of magnetic flux density that can be detected. Beyond LOD the sensor device noise signal dominates and the rms value of the output signal $b_{\text{out}}^{\text{rms}}$ is then less than the amplitude of the noise signal i.e., $b_{\text{out}}^{\text{rms}}$. LOD can be expressed as [13]:

$$\text{LOD} = \frac{1}{k}\sum_{i=0}^{K-1} b_{\text{out}}^{\text{rms}}(i) + 3\sqrt{\frac{\sum_{i=0}^{k-1}\left(b_{\text{out}}^{\text{rms}}(i) - \bar{b}_{\text{out}}\right)^2}{K-1}} = \mu_n + 3\sigma_n. \tag{6.2}$$

where, $K \in \mathbb{N}$, μ_n i.e., the first term is the mean value, and the second term i.e., σ_n is the standard deviation.

6.2.2 Current Sensor

Current sensing devices are utilized to manage an effective energy supply and to regulate or predict any fault in the electrical systems. There is an immense demand for highly safe and smart grids and therefore desirable to improve the common mode rejection rate by reducing the noise from sources including magnetic and electric field noise, low-frequency thermal noise, vibration noise. ME current sensor device comprises of a piezoelectric material upon which are fixed the two magnetostrictive plates on the side surfaces. The ME current sensor device detects ac current by observing for the presence of a magnetic field formed as per Ampere's law nearby a current carrying conductor and further converting the magnetic field to ME voltage as an output. However, since ME current sensors are subject to both electric and magnetic fields simultaneously, vibrational noise also become significant. Zhang et al. fabricated a ME current sensor device comprising of a pair of trilayer structure i.e., PZT sandwiched between two Terfenol D layers [14]. The pair of trilayer structure is separated by a distance of 16 mm biased by two NdFeB magnets of length 6 mm, width 3 mm, and height 2 mm as shown in Fig. 6.3.

A magnetic field gradient (MFG) is induced by the current carrying coil. The MFG in turn induces ME effect and we obtain an output ME voltage. Mathematically,

$$\text{MFG} = G = \frac{dB}{dr} = \frac{d}{dr}\left(\frac{\mu_o I}{2\pi r}\right) \tag{6.3}$$

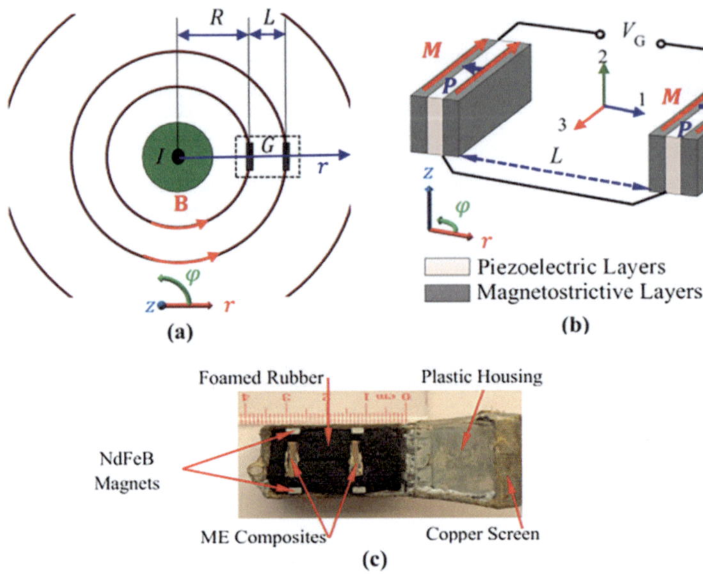

Fig. 6.3 **a** MFG induced by a current carrying coil, **b** represents ME (Terfenol D/PZT/Terfenol, D) current sensor comprising pair of two ME heterostructure separated by length "L". The output voltage is provided by the pair of ME structures, **c** original set-up. Adapted with permission from Ref. [14]

where, B is the magnetic field strength induced at a distance r from a current (I) carrying conductor/coil. Consequently, induced ME voltage (V_G) is then given as [14]:

$$V_G = \alpha_G G \tag{6.4}$$

where, G is the magnetic field gradient and α_G represents "gradient ME voltage coefficient". Also, the corresponding sensitivity is voltage change per unit current and can be expressed in terms of assembly parameters (R and L) as follows:

$$S_I = \frac{dV_G}{dI} = \frac{dV_G}{dG}\frac{dG}{dI} = \alpha_G \frac{d}{dI}\left(\frac{dB}{dr}\right) = \frac{\alpha_G \mu_O}{2\pi R(R+L)} \tag{6.5}$$

For Terfenol D/PZT/Terfenol D heterostructure ME current sensor, S_I of 0.62 mV/A corresponding to frequency range 1 Hz–80 kHz while S_I of 8.4 mV/A at resonant frequency of 120 kHz. In general, sensitivity value of 0.65–12.55 mV/A corresponding to frequency range 10 Hz–170 kHz with CMRR value of 17–28 dB.

6.2.3 ME Gyrator

The ME Gyrator is a two-port electrical network consisting of four key circuit elements namely, resistor, capacitor, inductor, and transformer [15]. It can interconvert voltage and current, and can convert network elements into their duals and vice versa, including inductor into capacitor, resistor into admittance, voltage source into current source. ME laminate structure composed of piezoelectric $PbZr_xTi_{1-x}O_3$ or $Pb(Mg_{1/3}Nb_{2/3})O_3$–30%$PbTiO_3$ sandwiched between two layers of magnetostrictive Terfenol-D in L-T mode [16]. A coil is wound around the laminate composite. This is observed to be a unidirectional gyrator with all the characteristics as aforementioned fulfilled. Also, a ME gyrator fabricated a copper coil wrapped trilayer structure comprising of the $Pb(Zr,Ti)O_3$ layer sandwiched between two ferrite layers of $Ni_{0.8}Zn_{0.2}Fe_2O_3$–0.3 wt% $MnCo_3$ and in L-T mode [17]. As shown in Fig. 6.4, the conducting contacts are attached from the two ferrite layers thereby forming a capacitive port. The entire ME core in enclosed by copper coil windings thereby forming an inductive port. The input current creates a magnetic field as per Ampere's law. The conversion of input power into magnetic power is then followed by stress–strain vibrations as a result of the transfer of magnetic power into the ME laminate core effectuated by direct ME coupling [16]. These stress–strain vibrations also result in the deformed piezoelectric layer which generates electrical output i.e., piezoelectric effect. The ME gyrator power transfer mechanism can be 'direct' which corresponds to the inductive port to capacitive port power transfer and vice versa for "converse" ME gyrator i.e., capacitive port to inductive port power transfer. For the aforementioned material, the power conversion efficiency (η) was found out to be 85% corresponding to power density (PD) of 20 mW/in^3 and η of 80% at PD of 5 W/in^3. In another study, the piezoelectric PZT layer sandwiched between two magnetostrictive layers of Metglas were reported to exhibit non-linear current/voltage (I/V) conversion coefficient of 0.59–1.56 V/A in low frequency (10 k Hz) and 43.16–116.47 V/A in the EMR region [18].

6.2.4 ME Transformers

A (2-1) ME laminate heterostructure comprising of a piezoelectric $Pb(Zr,Ti)O_3$ layer of thickness 100 μm is sandwiched between two magnetostrictive FeBSiC layers of 25 μm thick [19]. As depicted from the schematic, Fig. 6.5, an input ac current I_{in} is applied to the copper coils which are wounded over the ME laminate structure. This results in a magnetic field along the length of the ME laminate composite. The magnetic field induces a mechanical strain i.e., magnetostriction in the magnetic layer. This strain influences piezoelectric PZT layer and consequently we obtain an output ME voltage gain. The ME voltage gain is highest at the EMR because of the highest value of ME coupling at EMR. Notably, input dc component in current

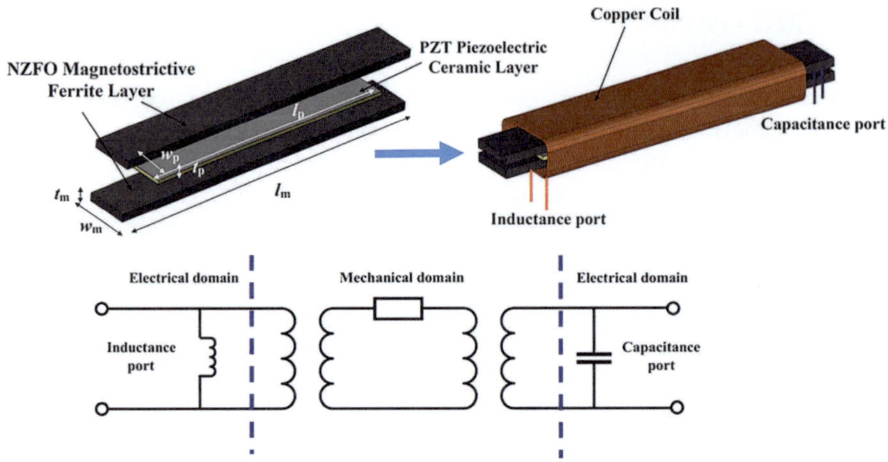

Fig. 6.4 Schematic and corresponding equivalent circuit diagrams for a ME gyrator consisting of a trilayer consisting of piezoelectric Pb(Zr,Ti)O_3 (length 30 mm, width 6 mm and thickness 0.5 mm) in between two magnetostrictive $Ni_{0.8}Zn_{0.2}Fe_2O_3$–0.3 wt% $MnCo_3$ layers of length 38 mm, width 5 mm and thickness 1 mm. Adapted with permission from Ref. [17]

also results in a dc magnetic field bias. This dc magnetic field bias controls magnetostriction and therefore the output ME voltage gain. This results in a tunable ME transformer.

Further, Fig. 6.6 represents an equivalent circuit model of a ME-based transformer (Fig. 6.5). V_{in} and I_{in} are the input voltage and current. I_{out} is the output current, V_{out} is the value of the output ME voltage obtained across load resistance R_L. Also, C_o represents clamped capacitance corresponding to the piezoelectric PZT layer and L^s represents clamped inductance corresponding to the coils wound around the ME laminates. Further, we define motional mechanical impedance (R_{mech}), motional mechanical inductance (L_{mech}), and motional mechanical capacitance (C_{mech}) as follows

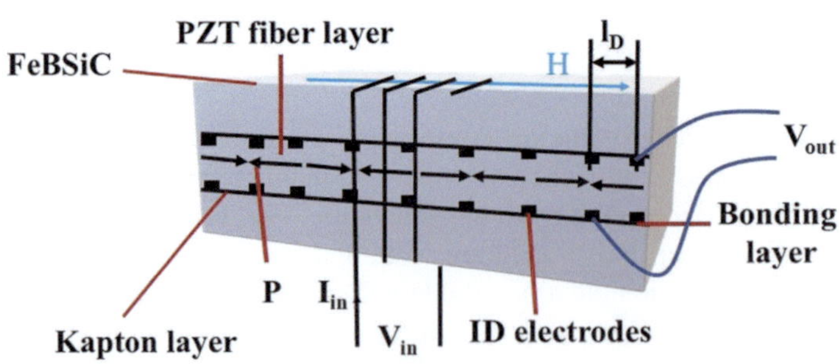

Fig. 6.5 Schematic of a ME transformer consisting of a PZT layer sandwiched between a '2-layer' metglas which is 0.36 mm thick and 12 mm wide and 100 mm long. Redraw in the impression of Ref. [19]

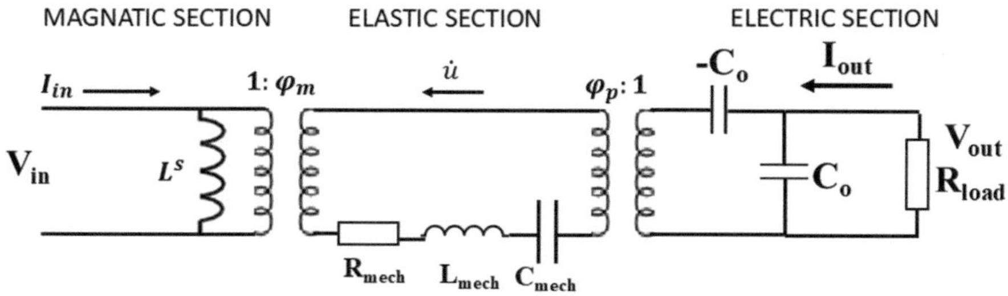

Fig. 6.6 Equivalent circuit representation of a transformer. Redraw in the impression of Ref. [20]

[20]:

$$R_{\text{mech}} = \frac{\pi Z_o}{4Q_{\text{mech}}},\ L_{\text{mech}} = \frac{\pi Z_o}{4\omega_s},\ \text{and}\ C_{\text{mech}} = \frac{1}{\omega_s^2 L_{\text{mech}}}. \tag{6.6}$$

Here, Q_{mech} is the mechanical quality factor corresponding to the piezoelectric layer, $\omega_s = 2\pi f_s$ where f_s is the resonance frequency and Z_o is the mechanical impedance corresponding to the ME laminate. The change in the magnetic phase induces mechanical deformation represented by a strain that can be expressed in terms of a magnetoelastic coupling coefficient (φ_m). The elastic deformation in the piezoelectric layer results in polarization in PZT and is represented by elasto-electric coupling (φ_p). Then, the maximum value of voltage gain and efficiency of the ME transformer can be expressed:

$$\text{VG}_{\text{max}} = \frac{4Q_{\text{mech}}\text{Ø}_p^2}{\pi\,\omega_s C_o Z_o}\ \text{and}\ \eta_{\text{max}} = \frac{\phi^2}{\phi^2 + \frac{\pi Z_o C_o \omega_s}{2Q_{\text{mech}}}} \tag{6.8}$$

It can be inferred from the above equation that higher value of quality factor gives increased ME voltage gain and therefore higher efficiency (η_{max}).

Further, Rosen-type piezoelectric (RTP) transformers have been utilized in order to obtain high output power density (step-up transformer). In addition, nowadays, in order to obtain a high ME output voltage, research studies employ magnetostrictive and piezoelectric materials. High permeability material $Fe_{73.5}Cu_1Nb_3Si_{13.5}B_9$ on Ni forms a magnetostrictive layer. This magnetostrictive layer is embedded on both sides of the drive end of RTP, $Pb(Zr_{1-x}Ti_x)O_3$ [21]. Similarly, another RTP is fabricated with the magnetic nickel–zinc ferrite ME transformer with a power efficiency of 35.8% at load resistance of 13 kΩ [22].

6.2.5 Wireless Power Transfer (WPT)

WPT has been used extensively in many modern-day applications including recharging devices, biomedical applications whereby an external power source can transmit power to an implant inside the body [23]. For instance, ME based transducers have been used effectively as wireless power receivers [24]. Biomedical usage of WPT technology can be classified into three categories namely, (i) radio frequency (RF) power transfer, (ii) acoustic transfer of power, and (iii) inductive power transfer. The first type i.e., RF power transfer has been used extensively in implants but is also prone to major disadvantage due to tissue heating at higher frequencies because of the high probability of absorption of radiation by the human tissue. In the acoustic WPT, the power is transmitted in the form of longitudinal i.e., sound/vibration waves. This power is then received at the other end by converting the sound/vibrational power into electrical power. Acoustic WPT is very safe for soft tissues, however, it is associated with considerable energy loss due to energy reflection at the air/tissue interface. Inductive power transfer (IPT) is commonly used for short distance and high-power wireless transfer. The disadvantage with IPT is that for very small size implants, the size of inductively coupled coils also becomes smaller thereby resulting in increase in the optimal driving frequency. The higher operating frequency will result in higher probability of energy absorption by soft tissues. Moreover, the inductive coupling coefficient reduces significantly as the distance between the transmitting and receiving coil increases. This reduces the efficiency of energy transmission. In general, for the requirement of a small size receiver, ME materials are good candidates for WPT systems. WPT involves two schemes and can occur through two schemes, (i) near field, and (ii) far field power transfer. Near field or non-radiative method of WPT utilizes a transmitter, a receiver, and for the transmission of energy, it is required to use either of the two flat capacitors (for energy transmission through electric field) or inductive coils (for energy transmission through magnetic field). These methods are preferred for bio-medical applications. The low power dissipation in lossy media implies more power transmission efficiency. On the other hand, far-field WPT involves RF electromagnetic (EM) waves for energy transmission. It can proceed further through microwave and laser WPT mechanisms.

ME transducer materials have been utilized as receiver for WPT. A 2-2 connectivity scheme ME-based (piezoelectric-magnetic) laminate structure can be fabricated in four types of modes as shown in Fig. 6.7. Here, "L-T" mode represents "Longitudinally magnetised-transversely poles", "T-T"" mode represents "transversely magnetised-transversely poled", "L-L" mode represents "longitudinally magnetised-longitudinally poled", "T-L" mode represents "transversely magnetized-longitudinally poled".

Figure 6.8a shows the WPT system comprising of a transmitter coil and a ME transducer as a receiver on the other end. The multiturn circular Helmholz coil is connected with a power amplifier and is used to generate the magnetic field. A PZT layer is sandwiched between the two Galfenol layers to form a ME transducer receiver. The layers of the ME transducer are held together using a conducting epoxy

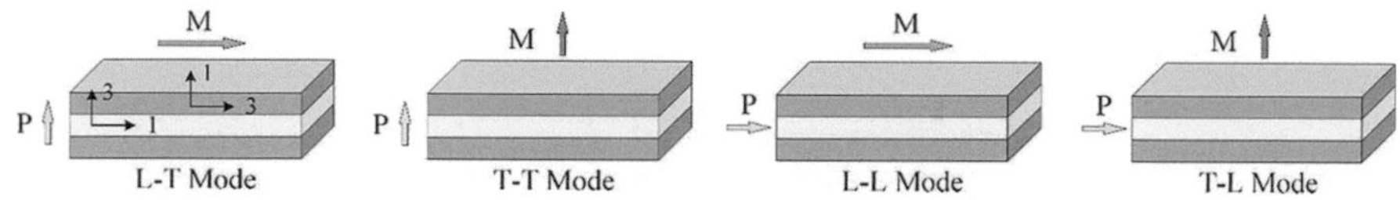

Fig. 6.7 Various modes corresponding to 2-2 connectivity scheme of ME laminate heterostructure. Piezoelectric material is sandwiched between two magnetostrictive materials. Adapted with permission form Ref. [23]

adhesive (EPO-TEK H20S). A DC magnetic field is provided as a bias to the ME transducer by positioning two neodymium magnets one above and one below the ME transducer. Figure 6.8b shows the equivalent circuit model. A uniform magnetic flux density is input to the receiver part of the WPT system. Mathematically we define, $F_o = \Gamma_m H_{ac} = \Gamma_m H_o \cos(\omega t)$, where, Γ_m is the magneto-elastic transduction factor. Thus, F_o represents the force that occurs due to the magnetostrictive layer. The applied magnetic field results in strain which in turn causes mechanical deformation in the magnetostrictive layer. Further, as observed in Fig. 6.8b, Z is the mechanical impedance, b is the damping coefficient. This also effectuates as a mechanical deformation in the piezoelectric layer where we define, Γ_p as the electromechanical transduction factor. In addition, C_o is defined as the clamped or electrical capacitance which corresponds to the piezoelectric material in between the two poles of the piezoelectric layer. Thus, the mechanical stress on the piezoelectric layer is responsible for generation of electric charge. Finally, the output voltage 'V' is obtained across the load resistance R_L.

Here, we have assumed ideal mechanical strain transfer from the magnetostrictive to the piezoelectric layer i.e., there is absence of any strain gradient at the interface. This can result in the overestimation of the ME coefficient and henceforth output ME voltage as we had disregarded the fact that the magnetostrictive and piezoelectric layers were fixed together using a conducting epoxy. However, we can experimentally determine the quality factor of ME laminate and incorporate into calculations corresponding to the equivalent circuit model.

6.3 Applications of Indirect ME Coupling

6.3.1 ME Random Access Memory (MERAM)

Magnetoresistive Random Access Memory (MRAM), a non-volatile and solid-state memory that is utilized to store binary data as 'stable magnetic states' and then to read data through measuring the change in resistance in the memory element. It comprises of a magnetic tunnel junction (MJT) composed of two magnetic electrodes which enclose an insulating tunnel barrier. Here, the resistance of the junction is a function of orientation of magnetic moment which determines the binary state "0"

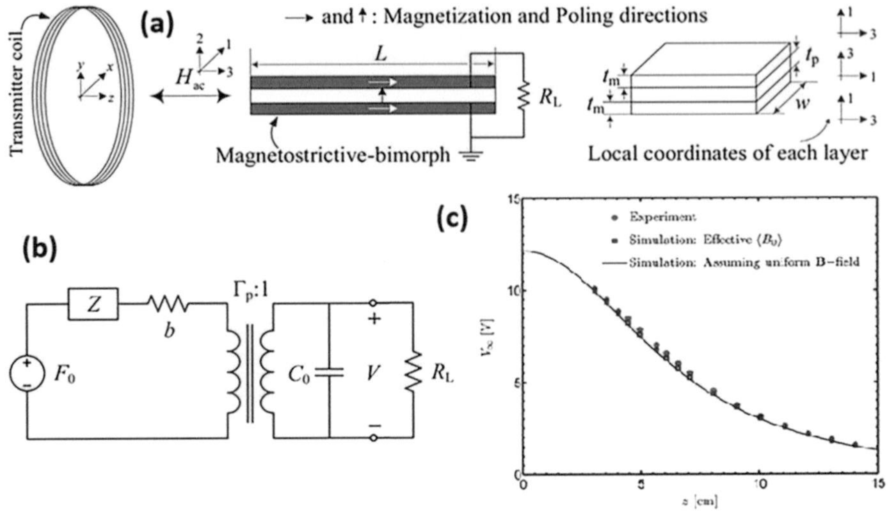

Fig. 6.8 **a** ME transducer as a receiver of a WPT system. (*x*, *y*, *z*) are global coordinates and (1, 2, 3) are local coordinates; **b** the equivalent circuit model of ME transducer as a receiver and **c** depicts the variation of output power when the ME transducer is moved away from the transmitter coil. Adapted with permission from Ref. [25]

and "1". Further, these magnetic bits can be read through detecting the change in the resistance of the junction. However, the writing of the magnetic bits requires high power because requirement of a high input current that is needed to obtain a high magnitude of magnetic field which is then utilized to store magnetic bits. In order to resolve this issue, the concept of "spin torque effect" is used whereby spin polarized current induces the magnetization in the material to switch directions. These are known as spin transfer torque MRAM. However, they still suffer from the disadvantage of high threshold current density. These limitations can be resolved using ME Random Access Memory (MERAM) [26–28]. MERAMs provide good energy efficiency i.e., the energy consumption gets reduced by a factor of 10,000 or even more. A MTJ comprises of two ferromagnetic (FM) layers separated by a thin insulating barrier (few nm thin where a spin current passes). The lower layer is the pinned layer while the upper layer is the free layer (magnetization direction is easy to rotate). The MTJ can be fabricated on a piezoelectric $Pb(Zn_{1/3}Nb_{2/3})O_3$–$PbTiO_3$ (PZN-PT) substrate [29]. Initially the magnetization of the two FM layers is in same direction (low resistance state). The electric field results in polarization and it induces deformations or mechanical strain which gets transported to the lower FM layer. The magnetic domains in this FM layer can reorient themselves. Thus, with the application of the electric field, the two magnetization directions in the two FM layers can become orthogonal to each other (high resistance state). In other words, the electric field can write "binary numbers" i.e., on a non-volatile "magnetoresistive memory". Figure 6.9 shows schematic of MERAM whereby the FE layer is PZN–PT, $BaTiO_3$ or $Pb(Zr,Ti)O_3$ with the FM "free" layer that can be $CoFe_2O_4$, Ni, Fe_3O_4

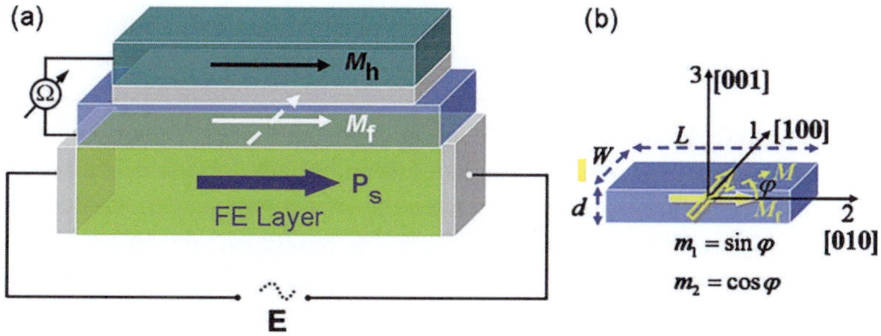

Fig. 6.9 **a** Schematic of a strain mediated MERAM. Here, M_h is magnetization in fixed/hard layer and M_f is magnetization in free layer, P_s is the spontaneous polarization induced due to applied transverse electric field, and **b** The orthogonal switching of magnetization in the free layer which is defined by its length, width, and thickness along [010], [100], and [001] respectively. Adapted with permission from Ref. [26]

and 001-oriented and/or a polycrystalline FeCo alloy. This provides low power, high energy efficiency and high-density memory elements. The ME-based memory writing does not impose any current-drive based restriction of size on the access devices like transistors in an IC. This allows for a smaller cell area as well. This enhances the binary bit density as well. Also, since the need for the current-operated device is reduced, the ohmic energy dissipation is also reduced.

Recently, antiferromagnetic (Cr_2O_3) MERAM (AF-MERAM) provides higher data stability and improves energy efficiency. Though the memory read operations are facilitated by the ferromagnet (FM) material, the presence of FM through the phenomena of hysteresis and exchange bias also interferes with the writing process in the memory [27].

6.3.2 *Microwave Tunable Filter Devices*

ME composites composed of the magnetic-FE phase material namely, $x\text{MnFe}_2\text{O}_4$–$(1-x)\text{Pb}_{0.988}(\text{Zr}_{0.52}\text{Ti}_{0.48})0_{.976}\text{Nb}_{0.024}\text{O}_3$ have found usage in the microwave tuning devices [30]. Microwave filter corresponds to the frequency range of 300 MHz–300 GHz. ME heterostructure comprising of a bilayer of PZT fixed with yttrium iron garnet (YIG) using an epoxy, is deposited on the gadolinium gallium garnet (GGG) substrate ((111) plane) [31]. This structure is then placed on the microstrip transducer connected to a transmission line with one end open and another shorted. The entire structure is then kept in between the poles of an electromagnet so as to get an applied bias magnetic field which is parallel to the plane of the sample film. The electric field upto 10 kV/cm is generated across piezoelectric PZT. Further, a continuous wave input signal of power 0.1 mW with frequency range 2–10 GHz was provided to the microstrip transducer. Here, the applied electric field is responsible for mechanical stress (converse piezoelectric effect) and henceforth deformation occurs which is then transferred to the magnetic ferrite layer using the ‘bundling effect’. The effective magnetic field on the magnetic layer constitutes the applied bias magnetic field, demagnetizing field and crystal anisotropy magnetic field. Then the change in the external applied electric field will result in a shift in the ferromagnetic resonance (FMR) frequency. This implies that the ME effect induces change in the FMR of the magnetic phase. This is also known as the microwave ME effect. This effectuates electric field tuning and consequently ME tunable bandpass microwave filter.

Figure 6.10 shows a ME microwave filter device comprising YIG deposited on the GGG substrate with PZN-PT forming a piezoelectric phase [32]. Further, insertion loss as a function of frequency varies for changing the magnetic and electric fields can be corroborated from the device as shown in Fig. 6.10. Figure 6.10 also helps to determine insertion loss as a function of frequency at the magnetic field of 150 Oe with varying width of the ME laminate structure. The maximum insertion loss of −1.588 dB is observed corresponding to the width of ME laminate of 3.6 mm with the −3 dB bandwidth of 47 MHz.

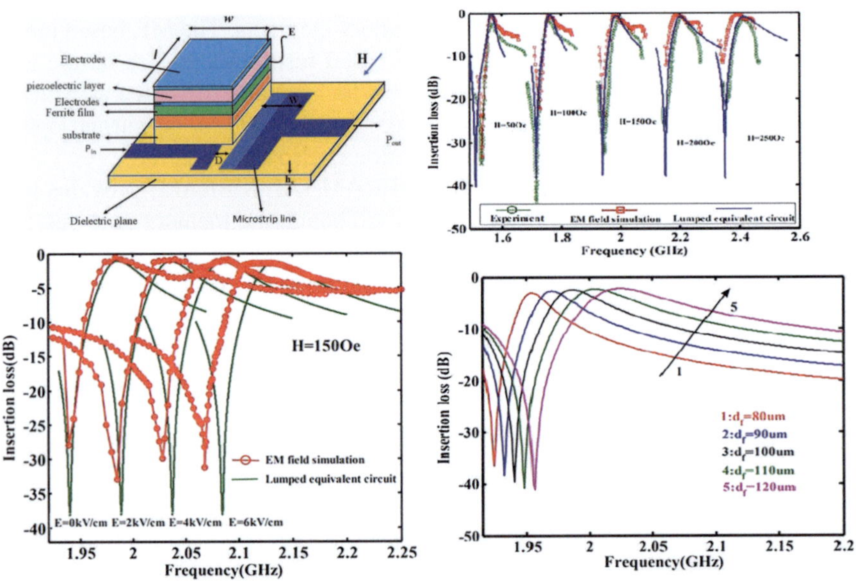

Fig. 6.10 Schematic of a ME tunable microwave filter to determine insertion loss as function of frequency for different values of magnetic field, insertion loss as function of frequency for different values of applied electric field on the piezoelectric layer, and insertion loss as function of frequency for varying thickness. Redraw in the impression of Ref. [32]

6.3.3 Phase Shifters

Phase shifters are utilized to change the phase of the electromagnetic signal (microfrequency range) in a controlled manner and find usage in the telecommunication system, phased array radars, oscillators etc. [33]. The phase shifters based on ferrite materials require high operating electrical power and henceforth their size cannot be reduced for use in miniaturized devices or integrated circuits. Further, phase shifters based on FE materials are effectively tuned by the electric field. But these exhibit lossy behavior at higher frequencies i.e., >1–5 GHz. Interestingly, fabricating a ME ferrite-FE layered structure can help to attain phase shift tunable by electric and magnetic fields. ME microwave phase shifters based on FMR comprise of magnetic ferrite (YIG) and FE (PZT) phase bonded together using an epoxy and deposited on a GGG substrate [34]. A silver electrode is taken from PZT which is poled whereby it is heated (150 °C) and cooled back to room temperature in the presence of an electric field of 20 kV/cm which is applied perpendicular to the ferrite/FE layer. This YIG-PZT ME resonator is placed on the microstrip transducer. The applied electric field on the piezoelectric PZT results in deformation due to mechanical stress. This establishes a magnetic field shift corresponding to the FMR frequency for YIG. Consequently, there is a phase shift observed for the propagating em waves in the microwave region. Notably, we can obtain a phase shift ranging from 90° to 180° corresponding to the voltage of 400 V applied across the piezoelectric PZT layer in the presence of an electric field of 8 kV/cm. Thus, the phase shifts (90°–180°) can be effectively tuned by the electric field with low loss and is compatible for miniaturized devices including complex integrated circuits. An electric field tunable ME layered phase shifter devices fabricated with M-type hexagonal $BaFe_{12}O_{19}$ (BaM) and piezoelectric PZT can provide differential phase shift of 45° with an applied electric field of 8 kV/cm [35]. The electric field tunable phase shift for the ME layered structure YIG-PZT is 45°–180° corresponding to FMR at 5–10 GHz and 45°–135° corresponding to magnetostatic surface waves at 6–9 GHz [36]. Similarly, ME bilayer structure comprising YIG grown on the GGG substrate with a barium strontium titanate i.e., $Ba_{0.5}Sr_{0.5}TiO_3$ (BST) grown on YIG using pulsed laser deposition method (PLD) was fabricated [37]. Further, on the top of the fabricated ME structure, coplanar waveguide (CPW) lines of varying lengths (1.5, 1.1, 0.25, and 0.2 mm) consisting of Cr/Ti/Ag/Au of thickness 50/25/2000/50 nm were deposited. The differential phase shift can be tuned successfully be magnetic field and electric fields. Electric field of 10 kV/cm was able to shift the FMR frequency by 5 MHz.

6.3.4 ME Voltage Tunable Inductors (ME-VTI)

ME materials find application in VTIs as well and often utilized in the circuit design [38–40]. VTIs comprise of magnetostrictive-piezoelectric layered materials as shown in Fig. 6.11a. Application of voltage induces mechanical stress in the piezoelectric

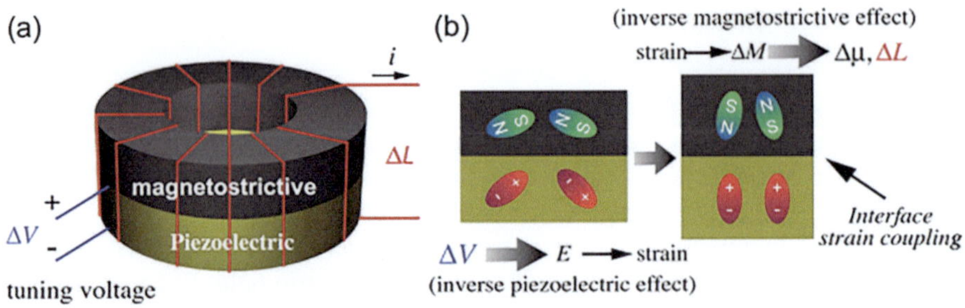

Fig. 6.11 **a** Schematic of a unit ME-VTI structure, **b** Mechanism of VTIs. Adapted with permission from Ref. [40]

layer which gets transferred to the interface region. The interfacial stress developed will influence the magnetostrictive layer by modifying the magnetic domains. Consequently, modulation of magnetic permeability is observed in the magnetostrictive layer. This is demonstrated in Fig. 6.11b, c.

The electric field tuned VTIs can be analyzed by the tunability relation as follows:

$$\gamma = \frac{K_\sigma}{K_o} = \frac{3}{2}\frac{\lambda_s}{k_o} Y d_{\text{eff}} E \tag{6.8}$$

where, γ is the tunability factor, K_σ is the stress anisotropy, K_o is the magnetocrystalline anisotropy, Y is Young's modulus, λ_s is the saturation magnetostriction coefficient, E the applied electric field, d_{eff} is the effective piezoelectric strain coefficient. It can be inferred from the above equation that for a good tunability in ME-based VTIs, we require K_o value to be low and λ_s to be high. Also, inductance tunability is defined as:

$$\gamma = \frac{\Delta L}{L_E} = \frac{L_O - L_E}{L_E} \tag{6.9}$$

where, L_O is the inductance in the absence of the electric field, and L_E is the inductance in the presence of the electric field. It is inferred from the experimental results that the inductance tunability depicts increasing trend with an increasing applied external electric field on the piezoelectric layer. In addition, inductance tunability exhibits stable behavior for frequencies up to 5 M Hz for PZT/Ni–Zn–Cu–Co ferrite thin layered ME-VTIs [38].

6.4 Acoustic Transmission and Surface Acoustic Waves (SAWs)

SAW transmission is defined as the propagation of the guided wave along the material surface whereby the wave vector is in a direction perpendicular to the surface with energy confined on the surface. The amplitude of SAW reduces exponentially as it is propagated down into the substrate. SAWs for the first time were observed by Lord Rayleigh in 1885 and find applications in identification of surface cracks, communication and signal processing, radio frequency signal filters, delay lines, bio-/chemical sensors, etc. [41]. In 1963, Rowen and Mortley, introduced planer surfaces to transducers that can be utilized to generate SAW as compared to working with the bulk waves. In 1965, the fabrication of the inter-digital transducer (IDT) by White and Voltmer saw generation of SAWs on the piezoelectric surface. SAWs find immense usage in the electromechanical systems for fabrication of sensing devices and signal processing [42].

6.4.1 Role of Strain-Mediated ME Coupling and Piezoelectric Effect in Surface Acoustic Waves

A SAW device based on transducer was fabricated using piezoelectric PZT films (0.2 μm thick) grown on (001) Nb:$SrTiO_3$ substrate [43]. A platinum electrode was deposited above the PZT layer. The piezoelectric materials are known to exhibit electromechanical coupling or generation of electric charge under application of mechanical stress. This electric charge is responsible for inducing voltage across the piezoelectric material. Further, converse piezoelectric effect is observed with an input applied electric field that generates a mechanical stress. The IDTs are patterned on the surface of the piezoelectric surface and are able to convert applied electrical energy into mechanical energy through the actuation process. This results in the generation of SAWs which propagate along the surface of the anisotropic piezoelectric substrate and are detected through the acoustic sensing. While propagating down the anisotropic medium, the SAWs are identified as completely longitudinal waves or transverse waves only along a certain specific direction. For instance, quasi-longitudinal are the waves exhibiting polarization very nearest to the propagation direction.

6.4.2 SAW Delay Line

A SAW delay line is as shown in Fig. 6.12 comprises of two IDTs separated with each other and patterned on the surface of a piezoelectric surface. Further, an AC input signal applied to first IDT effectuates converse piezoelectric effect which results in

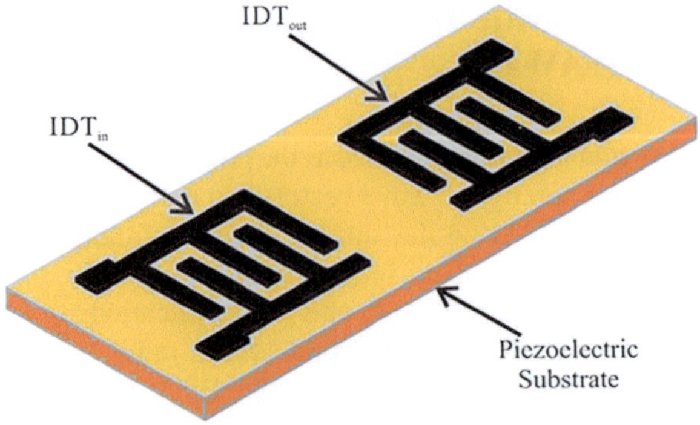

Fig. 6.12 Schematic of a SAW delay line. Here, IDT_{in} is input whereby an AC input signal is applied and output SAW is obtained from the IDT_{out}

the generation of SAWs. The generated SAW then propagates towards output IDT which is eventually converted back into electrical signal due to the piezoelectric effect. If 'L' is the distance between the two IDLs and V_o is the velocity of SAW, then the corresponding time delay in the signal propagation is given by $\tau = L/V_o$.

6.4.3 Acoustic Driven Magnetic Skyrmions

A novel spintronic device application is the study of magnetic skyrmions. A multilayer structure comprising magnetic-Ta(5 nm)/$Co_{20}Fe_{60}B_{20}$(CoFeB,1 nm)/MgO(1 nm)/Ta(2 nm) deposited on the piezoelectric $LiNbO_3$ substrate. The piezoelectric crystal orientation controls the propagation of SAWs. The SAW comprises of Rayleigh (RH) waves which possess both shear vertical displacement and shear longitudinal displacement, and shear horizontal (SH) waves which comprise of the shear longitudinal displacement alone. It is observed that RH waves are able to generate skyrmions but not move them while SH waves can also move skyrmions as a result of strong magnetoelastic coupling facilitated by in-plane strain gradients [44].

6.4.4 Acoustic Driven Magnetic Resonance

A coupling of SAW and magnetization of thin film effectuates acoustic ferromagnetic resonance (a-FMR) [45]. As referred from Fig. 6.12, the two IDTs are composed of Ti(5 nm)/Au(30 nm) and patterned on a piezoelectric $LiNbO_3$ substrate. Then the inverse piezoelectric effect drives the generation of SAWs of frequency 2.5 GHz whereas SAW are detected and reconverted to voltage signal by direct piezoelectric effect. The SAWs propagating on a ferromagnetic film (such as a Ni/Cu/Bi_2O_3 heterostructure) facilitate production of a time varying magnetic field as a result of a magnetoelastic coupling between time varying strain tensor field and magnetic domains. The RH wave component of SAWs comprises of a major longitudinal strain component whereby at a particular value of out of plane and plane angles, a minimum value of magnetic field, we obtain a maximum magnetoelastic strain. This results in a-FMR.

6.5 Conclusions

In this chapter, we have focused on multifunctional device applications of ME-based materials. It is important to highlight that the efficiency of MF- and ME-based devices is strongly dependent on the quality ME coupling. Consequently, optimization of ME coupling in the MF and ME materials also increases the efficiency of MF- and ME-based devices. It is observed that inorganic and lead-based materials such as $K_{0.5}Na_{0.5}NbO_3$, PMN-PT, PZT, PZN-PT, etc. are advantageous owing to their ease of fabrication, material properties, abundance, effective tuning with electronic devices. However, they are brittle and lead-based materials are hazardous to the environment. Thus, there is also an urgent need to develop and explore MF and ME materials exhibiting strong ME coupling at room temperature to further aid in the efforts towards manufacture of low cost, low power consuming MF and ME materials especially for SAW applications.

References

1. Gao J, Shen L, Wang Y, Gray D, Li J, Viehland D (2011) Enhanced sensitivity to direct current magnetic field changes in Metglas/Pb($Mg_{1/3}Nb_{2/3}$)O_3–$PbTiO_3$ laminates. J Appl Phys 109:074507
2. Gao J, Hasanyan D, Shen Y, Wang Y, Li JF, Viehland D (2012) J Appl Phys 112:10
3. Dong SX, Li JF, Viehland D (2004) Vortex magnetic field sensor based on ring-type magnetoelectric laminate. Appl Phys Lett 85:2307
4. Rupp T, Truong BD, Williams S, Roundy S (2019) Magnetoelectric transducer designs for use as wireless power receivers in wearable and implantable applications. Materials 12:512
5. Bichurin M, Petrov R, Sokolov O, Leontiev V, Kuts V, Kiselev D, Wang Y (2021) Magnetoelectric magnetic field sensors: a review. Sensors 21(18):6232
6. Dong S, Li J-F, Viehland D (2003) Ultrahigh magnetic field sensitivity in laminates of TERFENOL-D and Pb ($Mg_{1/3}Nb_{2/3}$) O_3–$PbTiO_3$ crystals. Appl Phys Lett 83:2265
7. Li M, Gao J, Wang Y, Gray D, Li J, Viehland D (2012) Enhancement in magnetic field sensitivity and reduction in equivalent magnetic noise by magnetoelectric laminate stacks. J Appl Phys 111:104504
8. Li M, Berry D, Das J, Gray D, Li J, Viehland D (2011) Enhanced sensitivity and reduced noise floor in magnetoelectric laminate sensors by an improved lamination process. J Am Ceram Soc 94:3738–3741
9. Li M, Matyushov A, Dong C, Chen H, Lin H, Nan T, Qian Z, Rinaldi M, Lin Y, Sun NX (2017) Ultra-sensitive NEMS mag-netoelectric sensor for picotesla DC magnetic field detection. Appl Phys Lett 110:143510
10. Krantz MC, Gerken M (2020) Limit of thermal-vibration noise in magnetic field detection with magnetoelectric-composite cantilevers. Phys Rev Appl 13:054047

11. Turutin AV, Kubasov IV, Kislyuk AM, Kuts VV, Malinkovich MD, Parkhomenko YuN, Sobolev NA (2022) Ultra-sensitive magnetoelectric sensors of magnetic fields for biomedical applications. Nanobiotechnol Rep 17:261–289
12. Özden M-Ö, Schmalz J, Gerken M (2023) A combined magnetoelectric sensor and human head model for biomagnetic FEM simulations. IEEE Sens J 23:30259–30270
13. Elzenheimer E, Bald C, Engelhardt E, Hoffmann J, Hayes P, Arbustini J, Bahr A, Quandt E, Höft M, Schmidt G (2022) Quantitative evaluation for magnetoelectric sensor systems in biomagnetic diagnostics. Sensors 22:1018
14. Zhang M, Wing OS (2018) Gradient-type magnetoelectric current sensor with strong multisource noise suppression. Sensors 18:588
15. Tellegen BD (1948) The gyrator, a new electric network element. Philips Res Rep 3:81–101
16. Zhai J, Li J, Dong S, Viehland D (2006) A quasi unidirectional… Tellegen gyrator. J Appl Phys 100:124509
17. Leung CM, Zhuang X, Friedrichs D, Li J, Erickson RW, Laletin V, Popov M, Srinivasan G, Viehland D (2017) Highly efficient solid state magnetoelectric gyrators. Appl Phys Lett 111:122904
18. Filippov D, Sapel'nikov S, Galkina T, Manicheva I, Poddubnaya N, Laletin V, Zhang J (2025) Magnetoelectric gyrator based on three-layer symmetrical structures: effect of magnetostrictive layer thickness. Sens Actuators A: Phys 388:116522
19. Dong S, Zhai J, Priya S, Li J-F, Viehland D (2009) Tunable features of magnetoelectric transformers. IEEE Trans Ultrason Ferroelectr Freq Control 56:6
20. Lu C, Li P, Wen Y, Yang A (2015) Large self-biased magnetoelectric properties in heterostructure of graded-magnetostrictive layers and a Rosen-type piezoelectric transformer. IEEE Sens J 15:1
21. Niu J, Li C, Xi Z (2024) The intrinsic piezoelectricity of monoclinic $Pb(Zr_{1-x}Ti_x)O_3$. RSC Adv 14(51):38245–38252. https://doi.org/10.1039/d4ra06270a
22. Zhang J et al (2024) Efficient power splitting and conversion in ferrite/piezoelectric transformer magnetoelectric composites. IEEE Trans Industr Electron 71:15179–15189
23. Saha O, Truong BD, Roundy S (2022) A review of wireless power transfer using magnetoelectric structures. Smart Mater Struct 31:113001
24. Truong BD (2020) Fundamental issues in magnetoelectric transducers: magnetic field sensing versus wireless power transfer systems. IEEE Sens J 20:5322–5328
25. Truong BD, Andersen E, Casados C, Roundy S (2020) Magnetoelectric wireless power transfer for biomedical implants: effects of non-uniform magnetic field, alignment and orientation. Sens Actuators A 316:112269
26. Hu J-M, Li Z, Wang J, Nan CW (2010) Electric-field control of strain-mediated magnetoelectric random access memory. J Appl Phys 107:093912
27. Kosub T, Kopte M, Hühne R, Appel P, Shields B, Maletinsky P, Hübner R, Liedke MO, Fassbender J, Schmidt OG, Makarov D (2016) Purely antiferromagnetic magnetoelectric random access memory. Nat Commun 8:13985
28. Wang S, Lee H, Ebrahimi F, Amiri PK, Wang KL, Gupta P (2016) Comparative evaluation of spin-transfer-torque and magnetoelectric random access memory. IEEE J Emerg Sel Top Circuits Syst 6(2):134–145
29. Pertsev NA, Kohlstedt H (2009) Appl Phys Lett 95:163503
30. Ciomaga CE, Avadanei OG, Dumitru I, Airimioaei M, Tascu S, Tufescu F, Mitoseriu L (2006) Engineering magnetoelectric composites towards application as tunable microwave filters. J Phys D: Appl Phys 49:125002
31. Fetisov YK, Srinivasan G (2006) Electric field tuning characteristics of a ferrite-piezoelectric microwave resonator. App Phys Lett 88:143503
32. Zhou H-M, Zhang Q-S, Lian J, Li X-H (2016) A lumped equivalent circuit model for symmetrical T-shaped microstrip magnetoelectric tunable microwave filters. IEEE Trans Magn 52(10):1–9
33. Cheng Y, Peng B, Hu Z, Zhou Z, Liu M (2018) Recent development and status of magnetoelectric materials and devices. Phys Lett A 382(41):3018–3025

34. Tatarenko AS, Srinivasan G, Bichurin MI (2006) Magnetoelectric microwave phase shifter. Appl Phys Lett 88:183507
35. Tatarenko AS, Srinivasan G (2011) A strain engineered voltage tunable millimeter-wave ferrite phase shifter. Microw Opt Technol Lett 53:261–264
36. Fetisov YK, Srinivasan G (2005) A ferrite/piezoelectric microwave phase shifter: studies on electric field tunabulity. Electron Lett 41:1066–1067
37. Leach JH, Liu H, Avrutin V, Rowe E, Özgür Ü, Morkoç H, Song Y-Y, Wu M (2010) Electrically and magnetically tunable phase shifters based on a barium strontium titanate-yttrium iron garnet layered structure. J Appl Phys 108:064106
38. Yan Y, Geng LD, Zhang L, Tu C, Sriramdas R, Liu H et al (2021) High-Power magnetoelectric voltage tunable inductors. IEEE Trans Ind Electron 68(6):5355–5365
39. Yan Y, Geng LD, Tan Y, Ma J, Zhang L, Sanghadasa M, Ngo K, Ghosh AW, Wang YU, Priya S (2018) Colossal tunability in high frequency magnetoelectric voltage tunable inductors. Nat Commun 9:4998
40. Yan Y, Geng LD, Zhang L, Gao X, Gollapudi S, Song H-C, Dong S, Sanghadasa M, Ngo K, Wang YU, Priya S (2017) Correlation between tunability and anisotropy in magnetoelectric voltage tunable inductor (VTI). Sci Rep 7:16008
41. Hess P (2002) Surface acoustic waves in materials science. Phys Today 55:42–47
42. Mandal D, Banerjee S (2022) Surface acoustic wave (SAW) sensors: physics. Mater Appl Sens 22:820
43. Kumar AKS, Paruch P, Triscone J-M, Daniau W, Ballandras S, Pellegrino L, Marré D, Tybell T (2004) High-frequency surface acoustic wave device based on thin-film piezoelectric interdigital transducers. Appl Phys Lett 85:1757
44. Yang Y, Zhao L, Yi D, Xu T, Chai Y, Zhang C, Jiang D, Ji Y, Hou D, Jiang W, Tang J, Yu P, Wu H, Nan T (2024) Acoustic-driven magnetic skyrmion motion. Nat Commun 15:1018
45. Puebla J, Xu M, Rana B, Yamamoto K, Maekawa S, Otani Y (2020) Acoustic ferromagnetic resonance and spin pumping induced by surface acoustic waves. J Phys D: Appl Phys 53:264002

Chapter 7
Conclusion and Future Outlook on Sustainable Multiferroic and Magnetoelectric Material-Based Technologies

7.1 Introduction

ME-MFs are a remarkable category of materials that simultaneously exhibit ferroelectric and magnetic properties, offering exciting opportunities for next-generation electronic devices, spintronic, and energy-efficient technologies [1, 2]. The synthesis methods and characterization techniques for ME-MFs are most important mainly focusing on advanced processes such as solid-state reactions, sol-gel, pulsed laser deposition (PLD), chemical vapor deposition (CVD), molecular beam epitaxy (MBE), and floating zone and green synthesis methods. Solid-state reactions yield complex oxide structures and chemical solution deposition techniques such as sol-gel process which enable the fabrication of high-quality thin films and nanostructures [3]. PLD is used for its precision in achieving epitaxial growth, critical for enhancing functional performance [4]. CVD is highlighted for its ability to produce high-quality thin films through chemical reactions in the vapor phase while MBE is noted for its precision in creating epitaxial layers in ultra-high vacuum conditions allowing for the fabrication of complex semiconductor structures [5, 6]. The floating zone method is discussed for its effectiveness in growing single crystals with minimal impurities and green synthesis process is emphasized for their environmentally friendly approaches to material production [7]. Characterization techniques are equally crucial for understanding the physical properties of these materials. X-ray diffraction (XRD) is employed to determine crystal structure and phase purity while scanning electron microscopy (SEM) and transmission electron microscopy (TEM) provide insights into surface morphology and internal structure at the nanoscale. Atomic force microscopy (AFM) is utilized for topographical analysis and impedance spectroscopy is essential for evaluating dielectric properties.

Owing to their remarkable success in photocatalytic applications, ME and MF materials have gained significant attention due to their exceptional coupling of electrical, magnetic, and optical properties. In addition to their appropriate band gaps, these materials exhibit inherent intrinsic polarizations enabling efficient charge

M. Kumar and S. Dabas, *Multiferroic and Magnetoelectric Materials*, Materials Horizons: From Nature to Nanomaterials, https://doi.org/10.1007/978-981-95-4391-5_7

carrier separation and their high mobility without the need for additional co-catalysts. The multifaceted applications of MF and ME materials are useful to emphasize their roles in energy harvesting for portable electronics [8]. Key developments include innovations such as self-charging MF modules, ultra-flexible ME heterostructures, and eco-friendly lead-free nanostructures that enhance energy storage and efficiency [8, 9]. The working principles of ME cantilever-based energy harvesters are useful in highlighting their resonance conditions for optimal performance [10, 11].

Also, the ME materials among the MF are emerging and appealing class of material with huge potential to be employed as photoactive materials for converting the photon energy into the electrical energy [12]. The co-existence of both ferroelectricity and magnetism simultaneously in the single and composite phase preferable at room temperature is the major factor for their application in the photovoltaic field. The coupling of the optical properties with the present ferroic property has made huge progress and revolutionizes the photovoltaic field. The bulk photovoltaic effect which is primarily observed in the ferroelectric material results in the smooth charge carrier separation based on remnant polarization which makes it possible to deliver the above band gap photo-voltage.

In conclusion, the chapter is majorly focused on the following points: (i) Explores ME coupling, dielectric constant measurements, and other dielectric properties providing a holistic view of the current advancements in the field of ME-MFs. (ii) The direct and/or converse ME coupling-based devices. The direct ME coupling-based devices are mainly the magnetic and current sensors, gyrators, transformers and wireless power transfer. And in case of converse ME coupling-based devices are mainly focussed as ME random access memory (MERAM), filters, phase shifters, and voltage-tunable inductors. (iii) Dedicated to the acoustic transmission, surface acoustic waves, the use of surface acoustic wave devices as radio frequency signal filters, delay lines, sensors, and acoustic-driven magnetic resonance. Also, the focus will be on the role of strain-mediated ME coupling and piezoelectric effect in surface acoustic waves. (iv) Devoted to the bio-medical applications such as drug delivery process, ME antenna for neurological applications, and ME coupling-based bio-magnetic field sensor. (v) Attempted to understand the Shockley–Queisser limit and quantum efficiency (internal and external) and the mechanism involved in the ME effect. Further, the possible green energy applications of the ME-MF materials also discussed. (vi) Try to discuss about the spin manipulation, magnetic exchange effect, strain-mediated ME coupling, prototype devices architecture, and spin-based transistors. The spintronic device-based technology will be the major focus. Figure 7.1 illustrates applications of ME-MF material-based technologies through the schematic diagram.

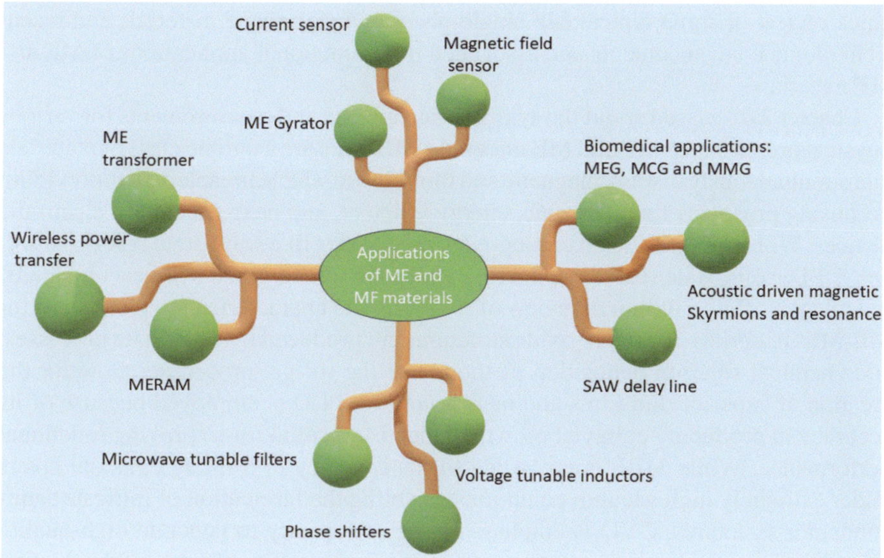

Fig. 7.1 Illustration depicting multifunctional device applications of ME and MF materials

7.2 Summary of the Present Work

MF-ME materials have garnered significant interest in recent years due to their unique ability to simultaneously exhibit multiple ferroic properties, including ferroelectricity, ferro/antiferromagnetism, and ferroelasticity. This multifunctionality positions them as promising candidates for a broad spectrum of advanced technologies such as spintronics, energy harvesting, sensing, and biomedical applications. One of the key advantages of the sustainable MF-ME materials is their exceptional structural tunability, which enables the integration of diverse functionalities within a single system. This adaptability supports the development of next-generation electronic and energy devices with enhanced performance and efficiency. This text presents a comprehensive overview of recent research and advancements in sustainable ME-MF materials. As far as the summary of this chapter is concerned, we have briefly discussed about the previous six chapters' summary.

In Chap. 1, the introduction part in terms of historical perspective and present development is discussed in detail. This chapter provides an overview of MFs, including their historical background and current state of development. Types of polarization and magnetization, Debye relaxation, and Cole–Cole plots are among the fundamental ideas introduced. We go into great detail into ferroelectric hysteresis, magnetic features including ferromagnetism and hysteresis, and antiferromagnetic materials. Direct, superexchange, double, and Dzyaloshinskii–Moriya interactions are among the classifications of exchange interactions. This chapter discussed type-I and type-II MFs, ME coupling, and an introduction to MFs. The chapter provides a quick review of some typical MF single-phase and composite materials and recent technological advancements and associated multifunctional applications of ME and MF materials.

Chapter 2 discussed about the synthesis techniques and measurements for various physical properties of MF and ME materials. ME-MFs are a unique class of materials that simultaneously display magnetic and ferroelectric (FE) characteristics, providing promising prospects for spintronic, energy-efficient, and next-generation electronic devices. With an emphasis on cutting-edge procedures like solid-state reactions, sol-gel, CVD, solid-state reactions, floating zone, MBE, and green synthesis methods, this chapter offers a thorough review of synthesis and characterization techniques for ME-MFs materials. Complex oxide structures are produced by solid-state processes, and chemical solution deposition methods like the sol-gel procedure allow for the creation of superior thin films and nanostructures. PLD is employed because of its accuracy in producing epitaxial growth, which is essential for improving functional performance. While MBE is renowned for its accuracy in forming epitaxial layers under extremely high vacuum conditions, enabling the fabrication of intricate semiconductor structures, CVD is emphasized for its capacity to generate high-quality thin films through chemical reactions in the vapor phase. The efficacy of the floating zone method in producing single crystals with few contaminants is examined, and the green synthesis process is highlighted for its ecologically beneficial methods of producing materials. Characterization methods are also essential for comprehending these materials' physical characteristics. While SEM and TEM offer information on surface shape and interior structure at the nanoscale, XRD is used to ascertain crystal structure and phase purity. Impedance spectroscopy is crucial for assessing dielectric characteristics and AFM is used for topographical analysis. The chapter offers a comprehensive overview of the most recent developments in the field of ME-MF materials research by examining ME coupling, dielectric constant measurements, and other dielectric properties.

Chapter 3 highlights sustainable MF and ME materials for energy harvesting applications. ME and MF materials have attracted a lot of attention because of their remarkable coupling of electrical, magnetic, and optical properties, which has led to their remarkable success in photocatalytic applications. These materials have the right band gaps and intrinsic polarizations that allow for effective charge carrier separation and high mobility without the use of extra co-catalysts. This chapter explores the various uses of MF and ME materials, highlighting their contributions to portable electronics energy harvesting. Self-charging MF modules, ultra-flexible ME heterostructures, and environmentally friendly lead-free nanostructures that improve energy storage and efficiency are examples of significant advancements. The concepts of ME cantilever-based energy harvesters are also examined in this chapter, with an emphasis on their resonance conditions for best results. With advances in self-powered systems and catalytic applications, the chapter delves deeper into the synergistic effects of integrating ME and photocatalytic processes for sustainable energy harvesting. Furthermore, the function of domain barriers in ME materials is examined, emphasizing how they affect magnetic coupling and polarization. This chapter summarizes the current obstacles and possible developments required to fully utilize the potential of these cutting-edge materials in next-generation energy solutions before offering suggestions for future research directions.

Chapter 4 succinctly covers photovoltaic performance of sustainable MF and ME materials in green energy technology. The ME and MF materials are a new and attractive family of materials that have a great deal of potential for use as photoactive materials that transform photon energy into electrical energy. The main reason for their use in the photovoltaic (PV) industry is that ferroelectricity and magnetism coexist simultaneously in single and composite phases that are preferred at room temperature. The combination of current ferroic characteristics with optical properties has revolutionized PV research and made significant strides in research toward next-generation photovoltaics and green applications. The bulk PV effect which is primarily observed in the ferroelectric materials results in the smooth charge carrier separation based on remnant polarization which makes it possible to deliver the above bandgap photo-voltage. The Shockley–Queisser (SQ) limit, internal and external quantum efficiency, and the mechanism underlying the ME effect were all important and comprehensively elaborated. Additionally, the ME-MF material's potential use in green energy was also explored.

Chapter 5 majorly covered ME-MFs for spintronic applications. This chapter clarifies spin manipulation and spin-dependent conduction. The spin-up and spin-down electrons act differently in an electric field. In trying to explain the resistivity of ferromagnetic metals at the Curie temperature, Mott originally suggested the significance of spin in 1936. Julliere originally identified tunnel magnetoresistance (TMR) in 1975 in Fe–Ge–Co junctions, which are found in Magnetic Tunnel Junctions (MTJ). Furthermore, the charge–spin coupling resulting from the spin–orbit interaction is the source of the spin Hall effect, which was initially theoretically predicted by Dyakonov and Perel in 1971. Giant magnetoresistance (GMR), which is seen in multiphase systems made up of alternating interlayers of ferromagnetic–nonmagnetic (conductive) materials, was then demonstrated. This chapter also goes into great detail into spintronic applications, such as spin transistors and the spin Hall effect.

Chapter 6 before this concluding remark this chapter is very important in highlighting the recent developments in ME coupling-based devices and applications in surface acoustic wave device technology. The direct and/or converse ME coupling-based devices are essentially the main subject of this chapter. The primary devices based on direct ME couplings include wireless power transfer, gyrators, transformers, and magnetic and current sensors. Conversely, ME coupling-based devices are mostly focused on filters, phase shifters, voltage-tunable inductors, and ME random access memory (MERAM). Additionally covered in this chapter are surface acoustic waves; acoustic transmission; and the application of surface acoustic wave devices as delay lines, sensors, acoustic-driven magnetic resonance, and filters for radio frequency signals. Additionally, the function of the piezoelectric effect and strain-mediated ME coupling in surface acoustic waves will be highlighted. This chapter continues the discussion of biomedical applications by focusing on medication administration, ME antennas for neurological applications, ME coupling-based bio-magnetic field sensors, and other biomedical applications. The important highlights of all the chapters are tabulated in Table 7.1.

Table 7.1 Highlights of the discussed six chapters of the present book

Chapters	Highlights
Chapter 1	• Basic concepts on ferroelectricity and magnetism. Introduction to MF, ME materials and ME coupling
Chapter 2	• Synthesis procedure and measurement techniques associated with ME and MFs • Comprehensive analysis on dielectric and impedance spectroscopy • Measurement and comparison of ME coupling coefficient
Chapter 3	• Integration of ME, MF and photocatalytic systems for green energy harvesting applications
Chapter 4	• The Shockley–Queisser limit and ME, MF-based photovoltaic devices for green energy applications
Chapter 5	• Role of ferromagnetism in evolution of magnetoresistance, TMR, MTJ devices, GMR technologies, and spin valve structures • Role of ME materials in spintronic applications
Chapter 6	• Review on applications associated with direct and indirect ME coupling mechanism • Role of strain-mediated ME coupling and piezoelectric effect in SAWs and their device applications

7.3 Future Outlook and Concluding Remarks

Over the past several years, the field of ME and MF materials has seen unprecedented interest in multifunctional device applications including sensing, memory devices, actuation, spintronics, etc. [13]. Majorly examined MFs are complex oxides which consist of at least two 3d transition metal cations with oxygen. Besides fluorides, organics and selenides are also classes of materials exhibiting MF properties [14]. Further, ME materials comprise composite system exhibiting simultaneously both ferroelectricity and magnetism. A major area of focus of current research in ME and MF materials has been the interfacial phenomena. This can be classified into two classes, one that arise spontaneously, i.e., domain-wall phenomena, and another is produced artificially through hetero-epitaxy technique. Specific orientation of domain walls, for instance, 109° and 180° for MF $BiFeO_3$, can be electrically conducting. These have potential to be utilized as information channels. Thus, the domain wall's orientation can be manipulated to switch between the conducting and insulating behaviors. Similarly, MF $YMnO_3$ exhibits MF vortex which is a topological defect associated with numerous intriguing phenomena and is a recent matter of research [15]. This has potential applications in nanoelectronics including memory storage or logic devices [16, 17]. Future of further research in the field of ME and MF materials lies in identifying materials and mechanisms that can be utilized to generate significant, consistent, robust room-temperature ME coupling at low values of applied electric and/or magnetic fields. Another major development is the emergence of two-dimensional MFs which include MOI_2 where "M" is Ti, V, Cr; trilayer NiI_2, etc., which are useful for advanced ME applications [18, 19]. Beyond the usually explored insulating and semiconductor MFs, researchers are also exploring "ferromagnetic polar materials" [20]. These materials are marked by intrinsic co-existence of structural polar distortions and magnetic orders and are associated with broken spatial-inversion and time-reversal symmetries [21]. This is similar to conventionally explored insulating MF materials. Ferromagnetic polar materials, therefore, provide a remarkable existence of ME coupling in a metallic system with applications in spintronic and ME devices. Further, in a quest for low power consumption along with high speed, i.e., "ps" memory devices, materials based on magnon, i.e., spin wave transport have also been explored [22]. The magnons can traverse through the magnetic materials without the actual movement of electrons. The field of ME and MF materials and their applications, though once assumed to be saturated decades back, continues to intrigue researchers worldwide owing to the exciting developments that have taken in the last few years. This field awaits numerous future developments with technological advancements that may arise on integrating MFs with other topological quantum materials.

References

1. Vopson MM (2015) Fundamentals of multiferroic materials and their possible applications. Crit Rev Solid State Mater Sci 40(4):223–250. https://doi.org/10.1080/10408436.2014.992584
2. Kumar M, Shankar S, Kumar A et al (2020) Progress in multiferroic and magnetoelectric materials: applications, opportunities and challenges. J Mater Sci: Mater Electron 31:19487–19510. https://doi.org/10.1007/s10854-020-04574-2
3. Lu C, Hu W, Tian Y, Wu T (2015) Multiferroic oxide thin films and heterostructures. Appl Phys Rev 2(2):021304. https://doi.org/10.1063/1.4921545
4. Hajlaoui T, Corbellini L, Harnagea C, Josse M, Pignolet A (2017) Enhanced ferroelectric properties in multiferroic epitaxial Ba2EuFeNb4O15 thin films grown by pulsed laser deposition. Mater Res Bull 87:186–192. https://doi.org/10.1016/j.materresbull.2016.11.033
5. Mantovan R, Vangelista S, Wiemer C, Lamperti A, Tallarida G, Chikoidze E, Dumont Y, Fanciulli M (2014) Synthesis of multiferroic Er–Fe–O thin films by atomic layer and chemical vapor deposition. J Appl Phys 115(17):17D907. https://doi.org/10.1063/1.4865774
6. Tabata H, Ueda K, Kawai T (1997) Formation of ferromagnetic/ferroelectric superlattices by a laser MBE and their electric & magnetic properties. MRS Online Proc Libr 494:201–212. https://doi.org/10.1557/PROC-494-201
7. Batulin R, Cherosov M, Kiiamov A, Zinnatullin A, Vagizov F, Tayurskii D, Yusupov R (2022) Synthesis and single crystal growth by floating zone technique of $FeCr_2O_4$ multiferroic spinel: its structure, composition, and magnetic properties. Magnetochemistry 8(8):86. https://doi.org/10.3390/magnetochemistry8080086
8. Hajra S, Vivekananthan V, Sahu M, Khandelwal G, Raj NP, Kim S-J (2021) Triboelectric nanogenerator using multiferroic materials: an approach for energy harvesting and self-powered magnetic field detection, Nano Energy 85:105964. https://doi.org/10.1016/j.nanoen.2021.105964
9. Aimon NM, Kim DH, Sun X, Ross CA (2015) Multiferroic behavior of templated $BiFeO_3$–$CoFe_2O_4$ self-assembled nanocomposites. ACS Appl Mater Interfaces 7(4):2263–2268. https://doi.org/10.1021/am506089c
10. Wu H, Tatarenko A, Bichurin MI, Wang Y (2021) A multiferroic module for biomechanical energy harvesting. Nano Energy 83:105777. https://doi.org/10.1016/j.nanoen.2021.105777
11. Onuta TD, Wang Y, Long CJ, Takeuchi I (2011) Energy harvesting properties of all-thin-film multiferroic cantilevers. Appl Phys Lett 99(20):203506. https://doi.org/10.1063/1.3662037

12. Onuta T-D, Wang Y, Long CJ, Lofland SE, Takeuchi I (2012) Dynamic state switching in nonlinear multiferroic cantilevers. Appl Phys Lett 101(4):043506. https://doi.org/10.1063/1.4738991
13. Srivastava V, Song Y, Bhatti K, James RD (2011) The direct conversion of heat to electricity using multiferroic alloys. Adv Energy Mater 1:97–104. https://doi.org/10.1002/aenm.201000048
14. Spaldin NA, Cheong SW, Ramesh R (2010) Multiferroics: past, present, and future. Phys Today 63(10):38–43. https://doi.org/10.1063/1.3502547
15. Spaldin NA, Ramesh R (2019) Advances in magnetoelectric multiferroics. Nat Mater 18:203–212. https://doi.org/10.1038/s41563-018-0275-2
16. Zhang Q, Tan G, Gu L et al (2013) Direct observation of multiferroic vortex domains in $YMnO_3$. Sci Rep 3:2741. https://doi.org/10.1038/srep02741
17. Li YH, Song YQ, Ren JH, Wang D, Liu K, Zhao ML, Ge J, Liao SJ, Xiao ZF, Li MM, Tian Y, Wang AJ (2025) Design of surface symmetry for low-voltage ferroelectric switching. J Appl Phys 138(2):024101. https://doi.org/10.1063/5.0278346
18. Xia Y, Yang X, Dou S, Cui H, Wei B, Liang B, Yan X (2024) Magnetization switching and performance of an optimized bicomponent multiferroic nanomagnet. AIP Adv 14(4):045239. https://doi.org/10.1063/5.0206692
19. Behera B, Sutar BC, Pradhan NR (2021) Recent progress on 2D ferroelectric and multiferroic materials, challenges, and opportunity. Emerg Mater 4:847–863. https://doi.org/10.1007/s42247-021-00223-4
20. Wu Y, Zeng Z, Lu H et al (2024) Coexistence of ferroelectricity and antiferroelectricity in 2D van der Waals multiferroic. Nat Commun 15:8616. https://doi.org/10.1038/s41467-024-53019-5
21. Zhang J, Shen S, Puggioni D et al (2024) A correlated ferromagnetic polar metal by design. Nat Mater 23:912–919. https://doi.org/10.1038/s41563-024-01856-6
22. Liu Z, Li Q, Chen H (2024) Ferromagnetic polar metals via epitaxial strain: a case study of $SrCoO_3$. Phys Rev B 110:024105. https://doi.org/10.1103/PhysRevB.110.024105